REALITY'S MIRROR: EXPLORING THE MATHEMATICS OF SYMMETRY

Other Wiley Science Editions

The Search for Extraterrestrial Intelligence, by Thomas R. McDonough

Seven Ideas that Shook the Universe, by Bryon D. Anderson and Nathan Spielberg

The Naturalist's Year, by Scott Camazine

The Urban Naturalist, by Steven D. Garber

Space: The Next Twenty-Five Years, by Thomas R. McDonough

The Body In Time, by Kenneth Jon Rose

Clouds in a Glass of Beer, by Craig Bohren

The Complete Book of Holograms, by Joseph Kasper and Steven Feller

The Scientific Companion, by Cesare Emiliani

Starsailing, by Louis Friedman

Mirror Matter, by Robert Forward and Joel Davis

Gravity's Lens, by Nathan Cohen

The Beauty of Light, by Ben Bova

Cognizers: Neural Networks and Machines that Think, by Colin Johnson and Chappell Brown

Inventing Reality: Physics as Language, by Bruce Gregory

The Complete Book of Superconductors, by David Wheeler and Mark Parish

Planets Beyond: Discovering the Outer Solar System, by Mark Littmann

The Oceans: A Book of Questions and Answers, by Don Groves

The Starry Room, by Fred Schaaf

The Weather Companion, by Gary Lockhart

To the Arctic: An Introduction to the Far Northern World, by Steven Young

Ozone Crisis: The 15-Year Evolution of a Sudden Global Emergency, by Sharon Roan

The Endangered Kingdom, by Roger DiSilvestro

Serendipity: Accidental Discoveries in Science, by Royston M. Roberts

Senses and Sensibilities, by Jillyn Smith

Chemistry for Every Kid, by Janice Pratt VanCleave

The Atomic Scientists: A Biographical History, by Henry Boorse, Lloyd Motz and Jefferson Weaver

The Starflight Handbook: A Pioneer's Guide to Interstellar Travel, by Eugene Mallove and Gregory Matloff

REALITY'S MIRROR: EXPLORING THE MATHEMATICS OF SYMMETRY

Bryan Bunch

Wiley Science Editions
John Wiley & Sons, Inc.

New York • Chichester • Brisbane • Toronto • Singapore

For my wife, Mary,
who maintains the symmetry
of my life.

Library of Congress Cataloging-in-Publication Data

Bunch, Bryan, H.
Reality's mirror : exploring the mathematics of symmetry / Bryan Bunch.
p. cm.
Bibliography: p.
ISBN 0-471-50127-1
1. Symmetry (Physics) I. Title.
QC174.17.S9B86 1989
539.7'25—dc20 89-33271
CIP

Printed in the United States of America

89 90 10 9 8 7 6 5 4 3 2 1

Preface

This book is for people who are curious about how the world works. *Reality's Mirror* attempts to explain the importance that mathematical symmetry has in modern science, especially physics. It also treats related topics in biology, art, anthropology, and music. With clarity and simplicity, it introduces a big chunk of interesting mathematics, along with some notions about how mathematicians work.

There are a number of good books that explain the impressive results that modern science has obtained through mathematics. Unless you are a mathematician, or well-acquainted with mathematics, many of these results might as well have been obtained by magic. How is it possible to predict a new particle that no one has ever observed, for example? This question, which haunts the last third of this book, can be answered by symmetry considerations, either explicitly or implicitly. I have tried to discuss the concept of mathematical symmetry in sufficient detail for the layperson to understand how it can lead to discovery. Along the way, I explain some games and puzzles based on symmetry, as well as a constellation of interesting facts related to symmetry questions. What I did *not* try to do was to develop mathematics or physics in detail, nor to derive major results of either subject.

I was initially drawn to the subject of symmetry by Martin Gardner's excellent book that is largely concerned with the opposite of symmetry *The Ambidextrous Universe,* which I read in its original edition in 1964 and its revision in 1979. Many specific examples of interesting asymmetries, as well as the idea of using games to illustrate symmetry concepts, were first introduced by Mr. Gardner, and I did not resist using some of these, although often in different contexts. My own research has also added to the storehouse of examples and ideas that can make science and mathematics meaningful to the general reader.

I assume throughout, as I did in *Mathematical Fallacies and Paradoxes,* that the reader has some experience with high-school algebra and geometry. In terms of mathematics alone (not counting the other topics), the first chapter explains line symmetry and briefly discusses how mathematics works as an axiomatic or abstract subject. Specific applications of these ideas are discussed in Chapter 2. Chapter 3 introduces additional mathematical ideas, including a brief look at transformational geometry in relation to symmetry, tesselations, and point symmetry. Chapter 4 shows how symmetry can be translated into arithmetic and algebra by means of the concept of parity. With this translation in hand, we further explore transformational geometry. Chapter 5 develops another aspect of mathematics, one less obviously related to symmetry, dimension. In Chapter 6, we show that symmetry can be used to make predictions about how the real world works. Chapter 7 is not about mathematics, but how the symmetry we have come to expect in the world fails in a few important ways. Chapter 8 introduces an important, but heavy, dose of mathematics: group theory. Group theory is generalized symmetry. In Chapter 9, symmetry in its most general sense is discussed which is, oddly enough, virtually the opposite of what you learned about as symmetry in school—that is, the symmetry of chaos.

That is the mathematical backbone of the book, but this book deliberately weaves together mathematics, science, and life in general. In addition to mathematics, the reader will encounter a little anthropology, some biology and chemistry, art and music, games and puzzles, modern particle physics, and even some astronomy.

I wish to express my gratitude to my wife, Mary, who read the entire manuscript and suggested many improvements in wording.

Bryan Bunch

Contents

REALITY'S MIRROR: EXPLORING THE MATHEMATICS OF SYMMETRY

1

A Moment for Reflection

Pooh is shown looking at his two front paws. "He knew that one of them was the right, and he knew that when you had decided which one of them was the right, then the other one was the left, but he never could remember how to begin."

A. A. Milne

"I'll tell you all my ideas about Looking-Glass House. First, there's the room you can see through the glass—that's just the same as our drawing-room, only the things go the other way. I can see all of it when I get upon a chair—all but the bit just behind the fireplace. . . . Well then, the books are something like our books, only the words go the wrong way. . . . How would you like to live in Looking-Glass House, Kitty?"

Lewis Carroll

Human beings find some arrangements more appealing than others. When people trim the natural shapes of trees and herbs to produce a pleasing landscape, they are looking for, among other things, balance. Putting all the tall delphiniums off center in a border or the large yews on one side of the walk and the small ones on the other just does not seem adequate. The same is true of ideas. When a philosopher, scientist, or mathematician puts ideas forward as something that will appeal to others in his or her field, it is generally considered good practice to make sure that the ideas have a kind of balance. The balance required, however, is not between large and small ideas; it is more complex yet simpler. Certain concepts or entities should have similar concepts or entities on the other side of the conceptual walk. Other concepts or entities can exist without requiring this balancing act. Why is this so? Let's look at an example.

A philosopher might think that since Being exists, Nonbeing must also exist; a scientist learns that for every action there is an equal and opposite reaction; a mathematician would be lost without negative numbers as well as positive ones. These are examples of concepts or entities that exhibit *balance.* The true, the good, and the beautiful all have their opposite partners.

Other concepts or entities, however, have no similar counterparts. For example, scientists have not found an opposite for mass—no entity that a scientist could label antimass. The need for balance is so strong, however, that some physicists predict that one day antimass will be found.

The nonexistent antimass should not be confused with the

existing *antimatter,* about which more will be said later. The prediction of the existence of antimatter is one of the most spectacular examples of using mathematics to describe some previously unknown entity. In science, the conclusion that certain entities must have matching, but opposite, partners is a direct outcome of interpreting a mathematical treatment of the entities in question. Antimatter emerges as one of as the consequences of equations that describe matter. Specifically, the equations that tell how the electron behaves imply the existence of a twin particle that is in some ways the direct opposite of the electron. When scientists found such a particle, the equations had already shown that it must be there.

The belief among scientists in this kind of balance goes even deeper. If some sort of balanced pairing does not exist, scientists may say that at some earlier time or in some other physical state the balance once did exist. By reaching for a higher energy, for example, it is predicted that the balance will be recaptured. This line of thinking has led to the development of huge machines that can reach such high energies and temporarily restore the balance. In turn, achievement of this goal has enabled scientists to begin to theorize how the universe began.

Some Things Look More Real Than Others

How does a ghost look? Not the kind of ghost that appears as just an ordinary person who is no longer among the living; rather the kind of ghost that presumably inspired the creator of *Casper, The Friendly Ghost.* That ghost and all the Halloween ghosts in white sheets are attempts to approximate the idea of a shapeless blob of ectoplasm. Such a blob does not appear to be real. In fact, nearly all objects in the real world have definite shapes. Compare the two blobs in Figure 1–1.

One blob is meaningless. The other one, which was made by folding the paper across the ink while the ink was still wet, suggests something in the real world.

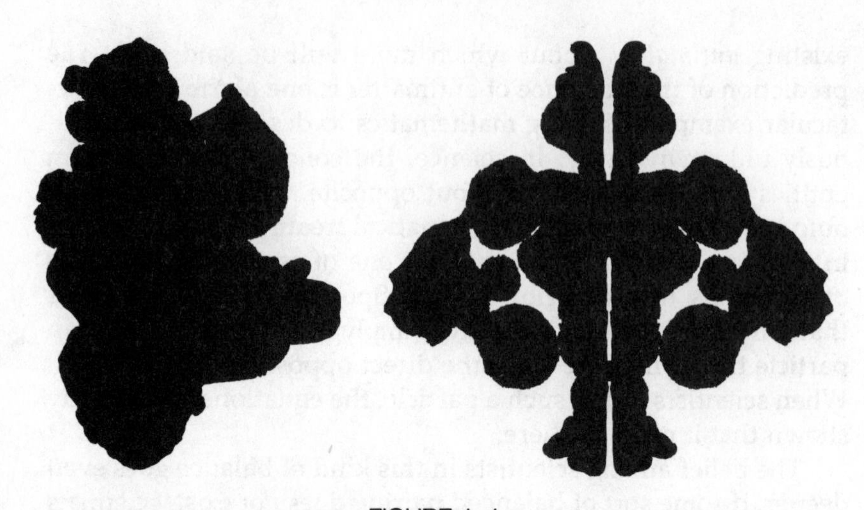

FIGURE 1–1
An ink blot that has been folded is generally more appealing than one that just spreads out any which way.

Psychologists use similar ink blots in a test for determining facets of the unconscious mind—the Rorschach test. In giving the Rorschach test, the psychologist presents the subject with a standard set of colored ink blots, one at a time. The subject describes what he or she sees in the blots. Some typical reactions to Rorschach's Card 1 (Figure 1–2) include three women doing a devilish dance, a witch perched on a rock being reflected from a still pool, an ash tray, some sort of bird, and a butterfly. Personally, I see a bat, perhaps a vampire. What do you see?

It would be easy to assume that all of the ink blots used in the Rorschach Test were chosen at random. In fact, just the opposite is the case. In the latter part of World War I in Switzerland, Hermann Rorschach, who was only one of several psychologists experimenting with ink blots at the time, spent several years making and choosing a set of ink blots that would unlock the inner feelings of subjects. Although Rorschach died (at the age of 37) shortly after publishing his work on ink blots and psychological testing, the Rorschach series has come to be

FIGURE 1–2
Some people see strange things in a Rorschach ink blot.

used universally. What characteristics do the Rorschach blots have that make them so effective? Perhaps it is a direct result of the process by which they were made.

The result of folding the paper over an ink blot is to produce a shape in which the left side is exactly like the right one. Most animals also share this characteristic, but most plants do not. This common aspect of animal and plant life is not just the result of chance. For the most part, animals move through the world and plants do not. Animals that do not move, such as sponges or oysters, lack a correspondence between right and left sides; they are more like plants, which have a tendency to look very much the same from whatever side they are seen. A rabbit has a well-defined right and left paw, but a tomato plant has right and left branches that are defined only by where you stand in relation to the plant. For this reason, the Rorschach blots are very often perceived as animals.

Plants evolved a way of getting energy from a relatively constant source that comes to them—sunlight. Animals, however, were forced to find sources of energy in various parts of the environment. Most animals move and most plants do not. A moving creature evolved a front and a back (with rare exceptions, such as starfish). It makes sense to have organs for seeing and feeding at the front where, as the animal moves forward, these organs encounter danger or food first.

Gravity is the primary determiner of the shapes of living things. Both plants and animals are affected by gravity, which constantly pulls from below. In fact, gravity determines what "below" is, as space travelers in free fall have found. One might then conclude that top and bottom are differentiated by gravity, and front and back are differentiated by motion. However, there is no force differentiating right and left. Therefore, right and left turn out to be the same (although reversed) on animals, although top and bottom, front and back are different. A plant, however, with no reason to differentiate from front and back, tends to grow to be roughly the same on all sides with only top and bottom different.

Why animals evolve to have right and left sides and plants do not is a (largely) nonmathematical example of an application of the idea of mathematical *symmetry.* "What can you possibly say about the subject after you've said that the part on one side looks just like the part on the other?" We will return to this question several times.

Animals have left and right symmetry, that is, one part is the mirror image of another part. This happens because the forces acting on the two parts are equal, but acting in opposite directions. When there is nothing acting on something to make a differentiation, most events that can happen will happen with about equal likelihood. It is this assumption that underlies the prediction from probability theory that a tossed coin will land with heads up about as often as it will land with tails up. If the coin were not symmetrical, the change in the coin's center of mass would cause it to land more often on the more massive side. In fact, people often put forth this idea to explain why the buttered side of a piece of bread seems always to land down when the buttered bread is dropped—although I have always preferred to believe that mystical forces in nature were at work in that particular case.

The result of the folding process used in making the Rorschach ink blots is to produce shapes that are always symmetrical. Therefore, when you see them displayed, it is easy to see animals (or humans) or even plants. The answer is a response to symmetry, not to a blob. Remember *The Blob,* a successful horror movie several years back? The emotional content of a blob is likely to be a simple negative reaction—xenophobia—since blobs do not look like the things we recognize as real. Another problem worth investigating, then, as we look at the role of symmetry in mathematics and science is whether or not the human aesthetic reaction to symmetry can sometimes influence our reactions to various kinds of experiences.

Don't Fold Now

At this juncture, it is useful to get a clearer explanation of mathematical symmetry.

A mathematical plane is just a very big, flat model of a piece of paper. By model I mean that the plane is not really a sheet of paper, but rather a mathematical idea that corresponds to a sheet of paper. Models are the way we go from the real world into the mathematical one and back. Another model we will be using is "3-space," technically Euclidean 3-space which is one of the mathematical models used for ordinary space with 3 dimensions—the physical space in which we usually think of ourselves as existing.

Now for another view of symmetry: If a mathematical figure is drawn on a mathematical plane, which has no thickness, and if that plane is folded—that is, one half the plane is rotated to meet the other half plane—through 3-space about a line, then the resulting figure in the other half-plane will be *symmetric* to the original figure. The ink blots were symmetric because they were made on paper that was folded in the real world in the same way as described for the plane in 3-space.

The definition of symmetry as rotation of a half-plane about a line in 3-space clearly shows that symmetric figures are congruent. Two mathematical figures are congruent if they have the same shape and size. Another way to say this is that reflection *preserves* shapes and sizes; that is, reflection preserves congruence. This notion of keeping certain relationships unchanged is one of the important ones of mathematics, and we will return to it in different settings using other words to describe it. The essential idea is that some relationships do not change under specified circumstances. For now, the only idea of this kind that you will have to use is that congruence for geometric figures is guaranteed if the figures are symmetric about a line. If the symmetric image is simply the original image that has been folded around a line, the size and shape of a figure will not change. Keep in mind that reflection in a mirror does not change size and shape.

The Art of Symmetry

The significance of symmetry to human beings goes back to prehistoric times. Just as the chimpanzee and the gorilla can be grouped into a larger family called apes, so can human beings be grouped with other species into a larger group called the *hominids.* As it happens, the only hominids around today (discounting legends of the yeti and bigfoot) are *Homo sapiens*—us. If you go back over the past 4 million years, however, other hominids show up in the fossil record. Two of these are known to be toolmakers, since the fossils are found in association with remains of tools. These are limited to tools that could survive several million years in the ground; that is, mostly tools made from rocks. Wood, leather, and most other materials do not have the staying power of rock.

Anthropologists believe that permanent tools are only made by hominids of the genus *Homo.* (Chimpanzees and sea otters, among other creatures, have been observed to make or use nonce tools.) Hominids called *Homo* are the ones that would pass the "subway test": If you shaved a hominid and dressed it in contemporary clothing, it can be called *Homo* if people would not think of the hominid as a strange creature when encountered on a New York subway. Hominids called *Australopithecus* are the ones that would fail the subway test. The Australopithecines are third cousins once, twice, or even three times removed.

The first known toolmaker is called "Handy Man," or *Homo habilis.* Don Johanson, who found the famous australopithecine fossil known as Lucy, commented that while *Homo habilis* would pass the subway test, if you saw one on the train, you would probably move to the other end of the car. Handy Man, who lived in Africa two million years ago, is still a long way from *Homo sapiens.* Handy Man got his name from the tools that were found in association with fossilized bones from his slight (four- to five-foot-tall) body. The tools are quite simple and just barely recognizable as artifacts. As a type, they are called "pebble tools." The types of tools found in association are called "toolkits," and are named after a site where they were found first or found in abundance. Pebble tools, first found in the Olduvai Gorge in East Africa, form the *Oldowan* toolkit. A collection of pebble tools is shown in the drawing in Figure 1–3.

The assumption that the tools were made by Handy Man is, perhaps, human egotism. Certainly some creature made and used the tools. For example, scratch patterns found on the fossil bones of creatures that might have been parts of the diet of

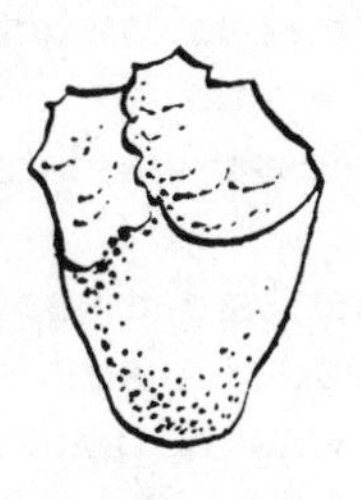

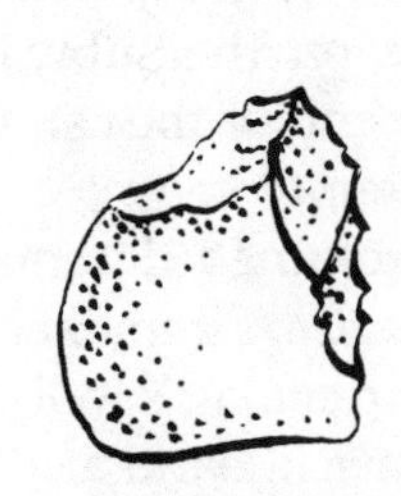

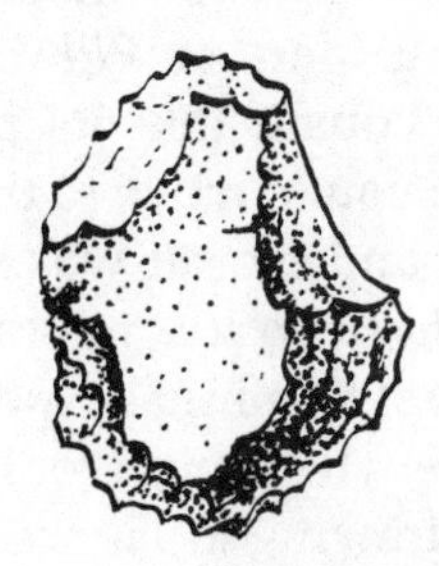

FIGURE 1–3
It is difficult to recognize that these chipped rocks were made on purpose.

Handy Man show that the carcasses had been butchered with pebble tools. While there are australopithecine candidates as the makers and users, other evidence suggests that it is, in fact, extremely likely that the tools were made by Handy Man, and that at least one of the purposes of the Oldowan toolkit was to remove the flesh from kills made by predators before *other* scavengers could get to it.

If you were walking through a gravel deposit, it would be very difficult to spot these tools unless you already had a good idea what primitive rock tools look like. In fact, when Louis and Mary Leakey first began to collect such tools in the 1960s, some members of the anthropological community denied that these were tools at all. The tools were derided as pebbles that had been chipped by natural forces. Eventually, computer studies of the way that the chipped edges were made established that the Leakeys were right: Some creature had actually made the tools. Recently, Nicholas Toth of the University of California, Berkeley, and the Institute of Human Origins, Berkeley, has been making such tools by randomly striking lumps of stone with a hammer. Although his tools are not quite so large or well-shaped as those of the first humans, they turn out to be effective in the same tasks—butchery, hide working, and woodworking.

These tools are difficult to identify as hominid creations because they do not show much in the way of symmetry.

Move forward a million years or so in time. Our second cousin *Homo habilis* has been replaced by our first cousin *Homo erectus,* perhaps more familiar to you as the Java Ape Man, Peking Man, or *Pithecanthropus erectus. Homo erectus* is such a close cousin that its presence on the subway would not even cause you to move to the other end of the car, unless it was late at night and the car was nearly empty.

Homo erectus might be nicknamed "Firemaker," since he was the first hominid known to use fire. Firemaker made tools as well as fire. The typical tool from Firemaker's toolkit was the hand ax. The drawing in Figure 1–4 shows a hand ax.

Firemaker did not change the basic design of the hand ax for a million and a half years. (In fact, the design was *never* changed

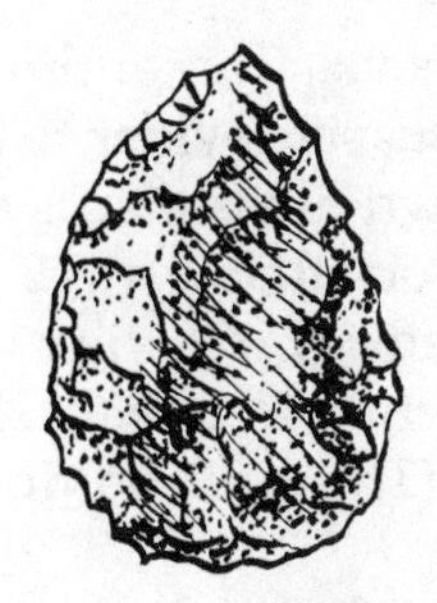

FIGURE 1-4
An Acheulean handaxe was clearly made on purpose, even if we don't know what the purpose was.

by Firemaker; it was replaced with the new and improved model at the same time that Firemaker was also replaced with a new and improved model of hominid.) Although Firemaker emigrated beyond Africa—at least to East Asia and Europe—the hand ax remained the same.

It is much easier to recognize a hand ax as an artifact than pebble tools. Toth, the modern toolmaker, says that these tools represent the first time that the stone knapper had a mental template that was being imposed on the raw material. One of the main reasons that the hand ax is instantly recognizable as a tool is that it has symmetry. While naturally occurring parts of plants and animals often have symmetry for good reasons, naturally occurring rocks (except for crystals, which are easy to recognize) do so only by accident. There could not be very many accidents that would produce hand axes, yet thousands of these rocks have been found all over the Eastern Hemisphere.

Why did Firemaker prefer a tool with symmetry? The first answer may be that such a tool is more effective than a tool without symmetry, but you should suspend judgment until more evidence is in.

For one thing, no one is entirely sure exactly what Firemaker used hand axes for. There are a lot of ideas, including butchering, but all of them are highly speculative. The people who first named the tool "hand ax" assumed that it was used for chopping as one would with a tomahawk. The anthropologist Eileen O'Brien has offered fairly persuasive evidence that the hand ax was a projectile that was thrown like a discus at animals gathered near water holes. The theory that it was a projectile fits in surprisingly well with William H. Calvin's theory that attributes the development of right-handedness, language, and a big brain to throwing rocks. (This will be discussed in more detail in Chapter 2.)

When *Homo sapiens* arrived on the scene, somewhere around the more recent ice ages and at least 300,000 years ago, we tossed out the Acheulean toolkit and, at a faster and faster pace, replaced it with a succession of toolkits of our own, currently with what we call the high tech toolkit. But 300,000 years before high tech, we were using what anthropologists call the Mousterian toolkit.

The people who invented the Mousterian toolkit, of which a tool called a point is shown in Figure 1–5, were the Neanderthals. Most anthropologists today recognize the Neanderthals as *Homo sapiens,* although a minority view is that they were a different species somewhere between Firemaker and us. There is fairly good evidence that they talked to each other, believed in gods and an afterlife, liked flowers, and fought with neighboring villages, all well-known signs of humanity. For our purposes, it is significant to note that our Neanderthal brothers and sisters preferred tools that are even more symmetrical than the tools Firemaker used.

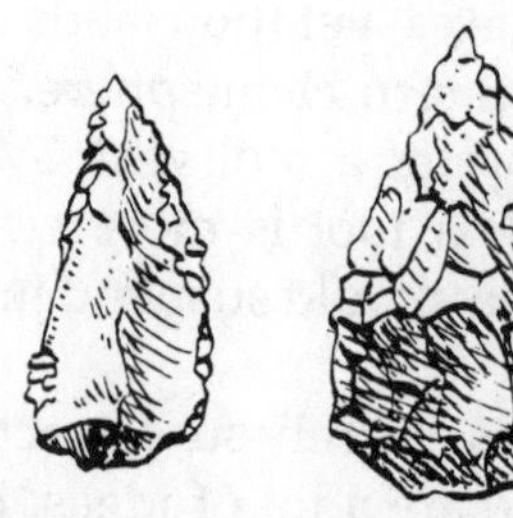

FIGURE 1–5
The Mousterian toolkit consists of tools that often show more symmetry than is really needed for their assumed purposes.

Neanderthal tools are in many cases enough like our own that we can be reasonably sure about the uses for which they were intended. A Mousterian point is useful as a spear head, for instance. Therefore, it is more profitable to wonder whether Neanderthal spear heads possess symmetry because of the known Neanderthal esthetic sensibility or because symmetrical spear heads are (1) easier to make, (2) easier to use, or (3) more effective. None of these latter reasons is sufficient. In addition to spearheads, Neanderthal people also made hand axes, but their versions are also much more symmetrical than earlier hand axes.

The principal consideration aside from esthetics in the manufacture of Mousterian tools is that a different technique was used to make them than the one used for earlier tools. The tools in the earlier or Acheulean kit were made by taking a piece of rock and chipping away on it until it was a tool. The Neanderthal toolmakers, however, started with a rock "core" and flaked off suitable pieces that were already part way toward being finished tools. Experiments performed by anthropologists show that the Mousterian manufacturing technique is more efficient than the earlier technique; however, the efficiency does not require that the tools produced be symmetrical (and, in fact, not all Mousterian tools are symmetrical; sometimes the core of rock used would not permit the manufacture of symmetrical tools). Symmetrical tools are not necessarily easier to make. In the case of points, it is not clear that the symmetrical points are either easier to use or more efficient. In fact, modern points (used for today's hunting arrows) are not always symmetrical. If this analysis is correct, the main reason for Neanderthal's preference for symmetry is esthetic; symmetrical points look better than asymmetrical ones.

This conclusion can be reached even more firmly with respect to the toolkits developed by the next wave of *Homo sapiens*, the people often called Cro-Magnon. By 20,000 to 30,000 years ago, the Cro-Magnons had carried the technology of making points from stone to its ultimate conclusion—the blade that anthropologists call "the laurel leaf." (See Figure 1–6.)

A laurel leaf might be as long as 11 inches and so thin as to be translucent. Most anthropologists believe that laurel-leaf

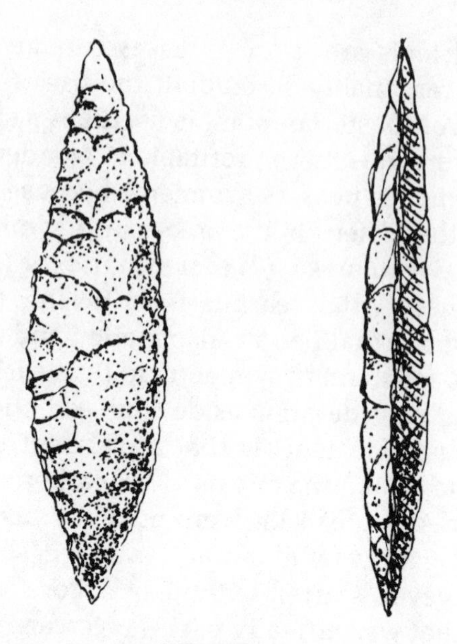

FIGURE 1–6
The laurel-leaf point was so delicate that anthropologists think it was made just for its beauty and not as a tool.

blades had no practical use whatsoever. They were simply works of art (possibly with some religious significance) and demonstrations of skilled artisanship. Laurel-leaf blades are also the first hominid-made objects to possess complete and perfect symmetry. The Cro-Magnons had developed an appreciation of symmetry for its own sake, an appreciation that seems to have been developing all along in the various hominid groups.

In their most well-known works of art, the cave paintings that came some 5,000 to 10,000 years after the laurel-leaf blades, the Cro-Magnons had advanced beyond simple symmetry in art to truly great art. Yet even then, geometrical patterns carved on bone or tusks reveal that the Cro-Magnons had not lost their love of symmetry. Symmetry is just too simple-minded for what they had in mind in making the great cave paintings.

From this time forward, you can find works of art that use symmetry directly and those that have balance without relying completely on pure symmetry. The use of pure symmetry, however, is generally relegated to architecture or the construction of tools, while more subtle forms of symmetry pervade works of fine art.

Don't Forget to Drop a Line

You may wonder what more you can possibly say about symmetry after you've said that symmetry means one side looks just like the other. But this is just one type of symmetry first encountered in elementary school and called simply, "symmetry." Since we will be dealing with more than one type of symmetry, it is better to receive a formal introduction to the concept using its full name—*line symmetry.*

While most people have a good general idea of what line symmetry is, in mathematics a good general idea is never enough. Mathematics requires that any topic in a system be defined in an unambiguous way.

Oddly enough, the way that these definitions are handled so that they are unambiguous is to make a kind of game out of them. Consider a game such as chess. It has a number of pieces that have familiar names, but which are really just undefined entities. The actual definition of the rook is determined by the rules of the game. There is no reason why a rook should be a piece of ivory or plastic shaped like an elephant or like a castle tower. If you lost the rook from your chess set, you could play the game perfectly well with a small pebble as long as you agreed that the pebble could only move vertically or horizontally and could only castle once a game (under specified circumstances). The game model for mathematics is used so often, in fact, that many mathematicians think that mathematics actually is a game with rules that mathematicians make up. As you learn more about symmetry, you can see for yourself how likely it is that this point of view is correct.

In mathematics, the ideas that correspond to the pieces in chess are called the "primitive" concepts of the game, and the game itself is called a "system." Like chess pieces, the primitive concepts cannot be defined apart from the rules of the game. To define the concept of line symmetry in strict mathematical terms, then, you need to know what primitive concepts will be used. It is sufficient to use *point, line,* and *ruler,* without knowing exactly what a point, a line, or a ruler is. Having agreed that these can be anything, mathematicians find it convenient to picture their undefined concepts in ways that are basically familiar. For a point, you can picture a small dot, rather like the period at the end of this sentence. Understand, however, that this is merely a convention, like showing chess pieces as small kings, queens, and elephants. You really have no idea what a point is; it could just as easily be pictured as an elephant. Similarly, for a line, you can picture a long, straight mark: ________; and you can think of a ruler as a meter stick. However, the rules of the game circumscribe the kind of image that is allowed. For example, a typical rule is: *A line is made out of points.*

In addition to the rules, it is necessary to have definitions. In chess, for example, it is necessary to have a definition of "capture" or "castle." For example, *the distance of a point from a line is the least number associated with the given point and any of the points that make up the given line, where the numbers are determined with a ruler.*

In formal systems, these definitions are fully worked out and based upon the simplest set of assumptions one can get away with. This definition, however, assumes that the rules of arithmetic are given and that such phrases as "associated with" have a clear meaning to the reader.

These rules and definitions are then combined to provide the following definition of symmetry: *Two points on different sides of a line are symmetric about the line if the distance between them is twice the distance of each of them to the line.*

This is not the ordinary definition of line symmetry, which relies on the notion of a "perpendicular bisector of a line segment," instead of distance and the somewhat nebulous "on different sides of a line." Since the distance from a point to a line is the length of a line segment from the point to the line that is perpendicular to the line, it should be clear that this definition means the same as the one that says the points are symmetric about the line if the segment that joins them is the perpendicular bisector.

The reason for basing a definition on distance is to relate it more directly to what you know to be true of the folded paper. The line is formed by the fold of the paper. The fold matches two points on the paper so that the distance from each to the fold is the same (and therefore the combined distance is twice that distance) (Figure 1–7).

A set of points is *line symmetric* to another set of points about a line if each point in the first set is symmetric to a point in the second set. The line is called a *line of symmetry.* The Rorschach ink blots are examples of line symmetry.

Although the definitions seem to deal with isolated sets of points reflected in a line, the same definition of line symmetry applies to any geometric figure, since a geometric figure is defined to be a set of points. Folding a paper can be used to see which geometric figures possess line symmetry, but that can be impractical. Another test with which you may be familiar uses a mirror (Figure 1–8). The mirror test can be used to show that some figures have line symmetry and others do not.

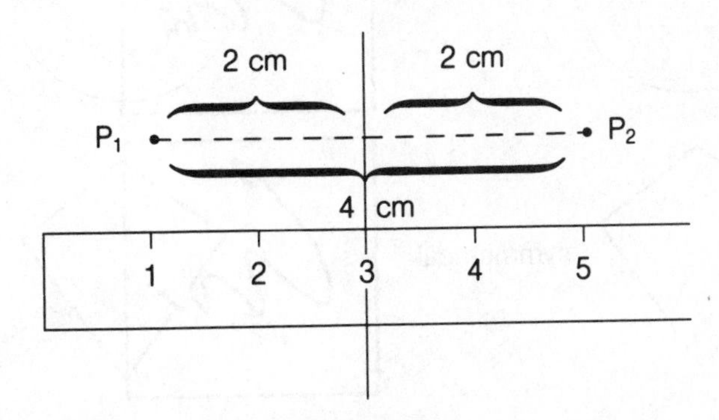

FIGURE 1–7
Two points are symmetric about a line if their distance to the same point on the line is half the distance between them.

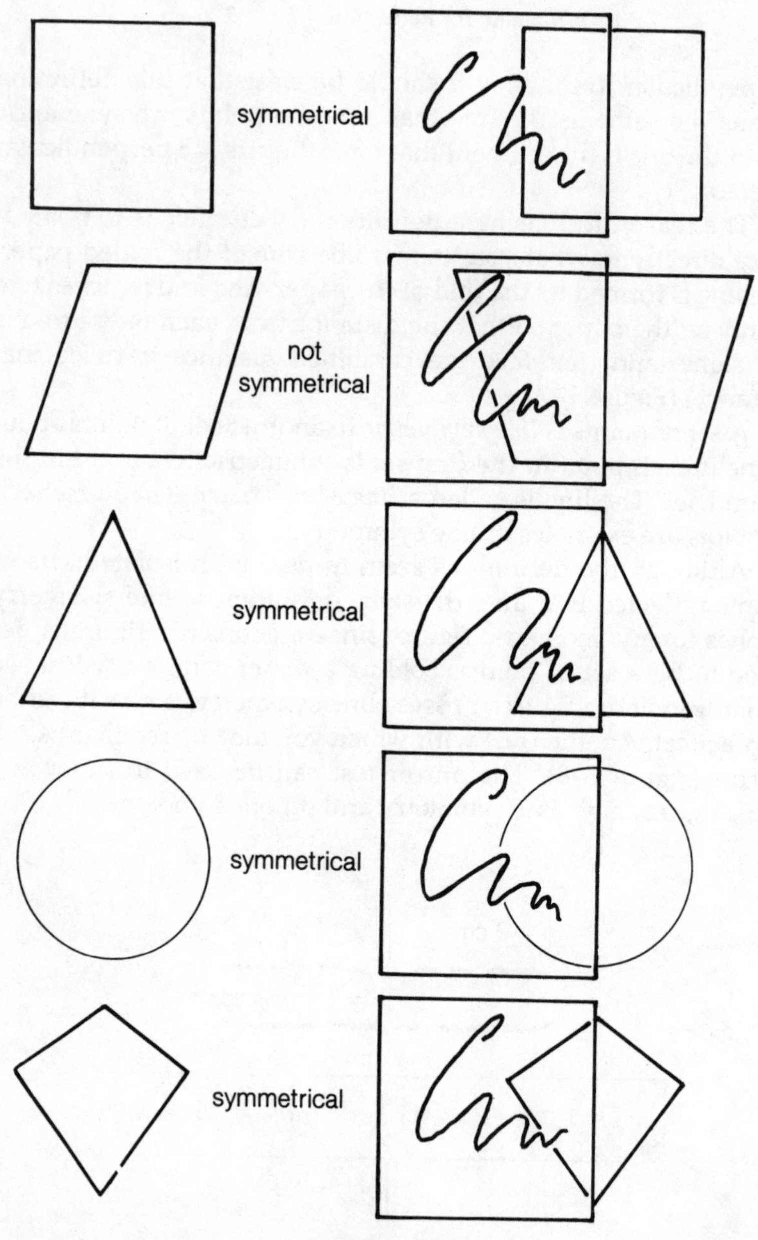

FIGURE 1–8
A mirror can be used to test whether or not a figure shows symmetry. The parallelogram (second from top) fails the test.

It is not always clear at first glance where the line of symmetry is. For example, in a sans-serif type (a typeface with no serifs or little tails on it) the following capital letters have line symmetry: A, B, C, D, E, H, I, K, M, O, T, U, V, W, X, Y. For example, A has a vertical line of symmetry down its middle, and B has a horizontal line of symmetry. Some of them have more than one line of symmetry such as H or O. It is a common exercise (which you may try for yourself) to find the lines of symmetry and to find the number of lines of symmetry for each letter. (Note: Some letters also have a property that you may recognize as symmetry, notably N, S, and Z. This property is not line symmetry, however. Develop a definition for this kind of symmetry using the same rules as were used in developing the definition of line symmetry.)

A reflection across the line of symmetry is just what the name implies. For ordinary symmetry with a straight line, a flat mirror placed along the line and perpendicular to the plane of the sets of points will not change the appearance of the total figure. The reflection of the points in the mirror will look exactly like the points that are behind the mirror.

Rounding Out the Idea

You may have noticed that there actually have been two different kinds of symmetry that have been discussed so far. Line symmetry can be identified by folding a sheet of paper or, more abstractly, by folding a plane. You cannot—in three dimensions at least—fold a substantial solid object such as a cube over on itself. Animals and stone tools, however, are usually substantial objects, and they exhibit a symmetry that is not quite line symmetry.

A mirror is not much of a guide to symmetry for solid objects, either. Although mirrors provide a useful way of checking the symmetry of figures drawn on a sheet of paper, the same approach is worthless for solid objects. When a mirror is used to check the symmetry of a figure in the plane, you hold the mirror in the middle of the figure and perpendicular to the plane. If you wanted to determine whether a cube is symmetrical, there is no convenient way to insert a mirror in the middle of the figure; and if you could do it, the mirror would not reflect the far side of the cube.

To get a better idea of how solid objects differ from plane objects with regard to symmetry, it is necessary to look back at the plane again.

We have agreed that two symmetric figures in a plane are congruent. For two dimensions, the image of congruence can be the same as Euclid's: Two figures are congruent if one can be superimposed on the other. This idea, however, can result in two slightly different interpretations of congruency. In one interpretation, the figures are merely moved about the paper (or plane) to achieve the superimposition. In that interpretation, the pair of figures on the left are congruent, but the pair on the right are not (Figure 1–9).

The pair of triangles on the right are not congruent in that interpretation because there is no way to slide one onto the other. That interpretation does not fit with our earlier ideas about folding planes or mirror images. The two triangles on the right *can* be made to coincide by folding the paper, for example. If you agree that symmetric figures are congruent, and it is clear that the pair of triangles on the right are symmetric, then you have to permit lifting the triangles out of the plane.

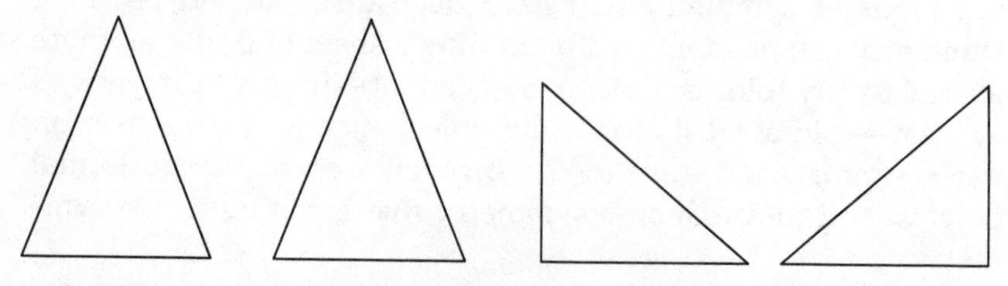

FIGURE 1–9
When an isosceles triangle is reflected in a mirror, as on the left, the resulting triangle is exactly the same as the original; but when a triangle that is not isosceles is reflected, the resulting triangle is somehow different from the original.

When it comes to three dimensions, however, the situation becomes more complex. Just as a mirror is physically unable to show the presence of symmetry in a three-dimensional object, it is also impossible to define three-dimensional congruence on the basis of superimposition and have a definition that reflects the situation for congruence in the plane. The most common example—at your fingertips, so to speak—is the pair of hands on a normal human being (Figure 1–10). Although the right hand should be congruent to the left hand to parallel the example of a right triangle that is congruent to a left triangle, no amount of twisting and turning can make the two hands coincide. The proof comes if you try to put a left glove on your right hand.

The two hands are congruent, however, but it is also worthwhile to have an additional word to describe the specific type of congruence. Mathematicians have coined the word, and it is a bit of a jawbreaker. The right hand is the *enantiomorph* of the left hand, and the left is the enantiomorph of the right.

There are many examples of enantiomorphic pairs that exist in the real world or that can be created. These range from parts

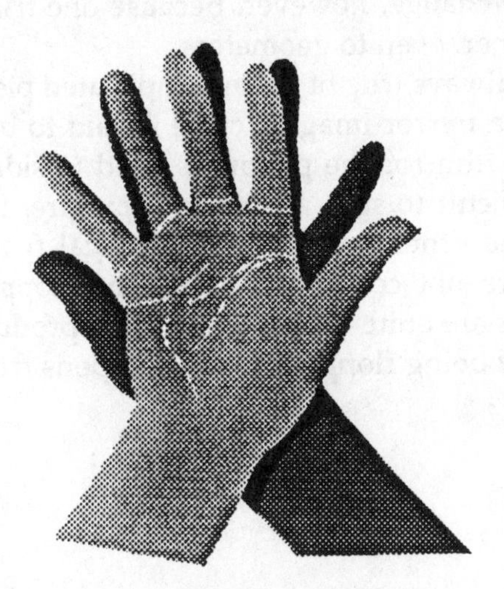

FIGURE 1–10
In three dimensions, the reflection is even less like the original, as left and right hands illustrate.

of the human body, such as the feet and ears, through knots, which can be tied in either left-handed or right-handed patterns, to that curious one-sided object, the Möbius strip, which can be formed with either a clockwise or a counterclockwise half twist. On the other hand, there exists a large group in the human population who find that not enough enantiomorphic pairs are available. This group is the left-handed, who are penalized in many sports (such as fishing with a rod and reel), in reading and writing (almost all books have the spine on the left), and frequently in work (if left-handed tools are not available).

This leads back to our original question about symmetry. "What can you possibly say about the subject after you've said that the part on one side looks just like part on the other side?" One thing to be said is that in three dimensions, the part on one side of a symmetrical object does not look just like the part on the other. It looks like its mirror image, which is always an enantiomorph of the real object.

This is true of two-dimensional objects, also. If a right triangle is shown in a mirror, the image is a left triangle. We are not sensitive to the change, however, because one triangle is pretty much like another, even to geometers.

This is not always true of more complicated pictures. In publishing jargon, a mirror-image picture is said to be "flopped." It happens when film for the picture is used upside down. Ordinarily, it is difficult to spot a "flopped" picture. Unless there is writing or some other notably asymmetrical feature showing, most people are not concerned about the flopped image (although painters are quite distressed to see reproductions of their work altered by being flopped, which happens frequently).

2

A Mirror to Life

Does nature have a way of defining right- or left-handedness uniquely?

Richard P. Feynman

She puzzled over this for some time, but at last a bright thought struck her. "Why, it's a Looking-glass book, of course! And, if I hold it up to a glass, the words will all go the right way again."

Lewis Carroll

The difference between right and left appears to be fundamental, but very hard to measure. The world we live in is a right-handed one, while Alice's looking-glass world was left-handed. That is because most humans are right-handed; so are our artifacts. At the banquet after Alice became a queen, the guests were all served wine in decanters, but had Alice needed to open an additional bottle of wine, she might have had some difficulty. The looking-glass corkscrew is an enantiomorph of the corkscrews we normally use. All helixes (spirals in three-space) are *asymmetrical.* The kind that we normally use are called right-handed and must be turned clockwise. A left-handed corkscrew will not work unless turned counterclockwise. If you want to cause someone great confusion, locate a left-handed corkscrew in a novelty shop. Most people will work at opening a bottle of wine with such a corkscrew for quite a while before they find out what trick has been played on them.

The Right Universe to Live In

The vast majority of people are right-handed. They prefer to use their right hands and, as a result, the majority of human right hands are stronger and more skillful than the left hands. Archaeologists and anthropologists confirm this human tendency to right-handedness over the ages, which is somewhat of a surprise.

Tools presumably made by our early ancestor *Homo habilis,* or "Handy Man," show right-handedness. Studies of later humans show that over 80 percent of the outlines of hands frequently found along with the famous Cro-Magnon cave paintings are of the left hand, suggesting that they were drawn by a right-handed painter. Similarly, an analysis of which hand was holding the weapon used by early man to attack baboons shows about 90 percent of the attacks resulted in bone damage that could only be caused by a right hand.

Our nearest relatives, however, the chimpanzees and gorillas, do not show consistent handedness. Instead, like all other mammals studied, roughly 50 percent of great apes are left-handed and 50 percent are right-handed. An individual mammal does tend to stick to the same handedness for the same tasks. For example, a dog that has been taught to shake hands with its paw will always use the same paw. But about half of all dogs studied prefer the left paw and the other half prefer the right.

The tendency for being right-handed is so ingrained in humans, however, that even the word for not favoring either hand, *ambidextrous,* actually means both right hands. Words that mean left hand have a definite negative connotation; *sinister* and *gauche* both are derived from words that mean left. *Adroit* and *dexterity* come from words that mean right. One has to suspect that the unknown Frenchman who divided political parties into "right" and "left" was a rightist trying to downgrade the opposition, although tradition has it that the division came from the seating arrangements in the legislative chamber. But even the seating suggests some prejudice against the left, for do we not seat the guest of honor on the right?

In English we write and read from left to right, but this is hardly a major human tendency. Semitic peoples generally write and read from right to left, although they are just as right-handed as the rest of us. The early Greeks had an economical system of writing called *boustrophedon* in which the lines alternated from left to right and from right to left. High-speed impact printers today do the same thing, but they print the right-to-left line *backwards,* so it reads from left to right. Boustrophedon writers expected to have boustrophedon readers, for their right-to-left lines also read from right to left. *Boustrophedon,* by the way, means "like an ox plows," a good description of the writing and reading method.

William H. Calvin's theory about the importance to human evolution of throwing stones was mentioned in Chapter 1. He believes that the Old-Worldwide expansion of Firemaker—the first hominid to leave the tropics—was due to the development of the ability to throw stones successfully at small game, producing meat for the table. In the tropics, there is enough of a year-round fruit and vegetable supply to keep a hominid alive, but outside the tropics, fruits and vegetables are seasonal. Success at killing small game could keep the troop alive during the winter, for example.

How would the development of throwing influence handedness? It would seem that ambidextrous throwing would be more beneficial than specializing in a single hand. The title of Calvin's book on the subject—*The Throwing Madonna*—suggests his concept, which I must admit I find a little far fetched. Calvin hypothesizes that women often hunted small game while carrying an infant in their arms. One arm would be used for the infant and the other for throwing. Since the human heart can be heard more loudly on the left side of the body, the baby would be lulled to sleep by being carried with the left arm, leaving the right for throwing. Evolution would strongly favor this, since crying babies would frighten game (and attract large predators). Well, I said it was a little far fetched.

Calvin's theory is not the only one that attempts to account for why humans have consistent handedness and other mammals do not. In the nineteenth century, it was suggested that the weight and position of the inner organs in human beings tended to make bodies list to the right. Compensating for this, we should balance better on the left foot, which would make the right hand freer than the left. Since nearly all humans have the same arrangement of inner organs, this theory does not explain why as many as 10 to 20 percent of all humans are left-handed. Thomas Carlyle, among others, held that the development of the shield was crucial to the issue. Since the shield would be held in the left hand to protect the heart, the right hand would be the active fighting hand. In addition to not explaining all those left-handers, the Carlyle theory does not hold up in light of the accumulated evidence that humans were mostly right-handed a million years before the shield was first used.

As will become clearer, handedness in humans actually seems to result from the brain, not from the muscles. Experiments designed to show what causes some brains to be different from others have yielded inconclusive results. Some studies have shown that a purely genetic determination along the Mendelian lines is unlikely. There is not enough evidence to indicate that the home environment of a child makes much of a difference, although parents and schools have had mixed success in teaching natural left-handers to be right-handed for some activities. The most likely explanation (I think; not everyone would agree) involves the prenatal environment. We will go into some detail on how this might work later in the chapter. It should be noted, however, that even this explanation does not explain why consistent handedness appears in humans and not in other mammals.

Animal Righties and Lefties

Although only human beings are predominately right-handed, consistent rightness and leftness are not unknown in nonhuman creatures. The way that this tendency is expressed, however, is quite different from the handedness shown by humans. One of the most remarkable examples is the tusk of the small arctic whale, the narwhal. This tusk grows in the male from one of only two teeth that are fully developed. (In rare instances, both teeth may grow into tusks.) The single tusk, which may be 7 or 8 feet long, always develops from the left tooth. Furthermore, as it grows, the tusk forms grooves that spiral round the tusk to form a left-handed helix. One might expect that if two tusks developed, the one on the right side would grow grooves that form a right-handed helix, but this is not the case. Both tusks are left-handed. No one has offered a very convincing explanation of why this should be so (nor has anyone offered much of a reason for most humans to be right-handed). Furthermore, no one is really sure what the male narwhal uses its tusk for. Most scientists rely on that old Darwinian catch-all, sexual selection. If that is the case, it would imply that the female narwhal finds left-handed spirals more attractive than right-handed ones. Biologists do not appear to have studied this.

Another natural example of right and left symmetry concerns certain tiny shelled creatures, related distantly to paramecia and diatoms, called *Foraminifera.* Foraminifera live in the oceans of the world. Their shells either coil to the left or to the right. Scientists noticed that the direction of the coil on the shell of a particular species of Foraminifera was dependent on the temperature of the part of the ocean in which it lives. If the temperature of the water is less than about 7 degrees Celsius (about 45 degrees Fahrenheit), the shells coil to the left. If the temperature is higher, the shells coil to the right. (Other related species also show this tendency, although not so dramatically.) No one knows why this happens. Since Foraminifera have been around for millions of years and since the shells are often preserved in the sediments at the bottom of the ocean, the direction of coiling can be used to tell something about the temperature of the ocean in the past. When used with other data about the age of sediments recovered from deep below the ocean's floor, Foraminifera have been used to provide unusually accurate data on the onset and termination of the most recent waves of ice during the Ice Ages.

Why such creatures vary from left-handed to right-handed is often quite mysterious. One investigation has been conducted into spiralling in a snail. The researcher, C. P. Raven of the University of Utrecht, found that the direction of shell coiling in a particular snail is determined by the genes of the snail's mother, with the father having little or no genetic influence. However, the mother does not possess a gene that is directly concerned with shell coiling. Instead, the mother's gene determines the shape of the egg. If the egg bulges a little in one direction, the resulting snail will coil to the right. If the egg bulges a little in the other direction, the snail will coil to the left. The side of the bulge determines the place where the egg will make its first division into two cells. Which of the two cells gets the larger amount of cytoplasm from the egg determines the direction of coiling.

Another extensive investigation concerned flounders. Flounders and their relatives, the sole, the turbot, the plaice, and the halibut (among others), are peculiar fish. When a flounder first hatches, it swims through the water just as other fish do. Its eyes are on different sides of its head, as in other fish as well. When the flounder gets to be a certain size, however, it lies down in the mud. Some have said that the flounder is just plain lazy. When it settles in the mud, it lies on its side. To enable it to see with both eyes in this situation, one of the eyes migrates to the other side of the head. As a result, both eyes are on top, looking up, away from the mud.

It has often been thought that one eye migrates as a result of settling into the mud. One assumes that there is an equal chance of the flounder lying right-side-down or left-side-down. However, the eyes often migrate before the flounder settles down into the mud. Therefore, it is unlikely that chance determines whether the flounder will be right or left side-down.

In some species of flounder, almost all of the fish are right-eyed, while in others, almost all are left-eyed. The starry flounder of the Pacific, however, presents a particularly interesting case. Near the West Coast of the United States, about half of the flounder are right-eyed and half are left-eyed; near the coast of Japan almost all of the flounder are left-eyed.

David Policansky of Harvard tried to find out why. He thought there might be something in the environment of the Western Pacific that favored left-eyedness. It is hard to imagine what this might be, however, since nature seems so incessantly symmetrical. His studies suggested that left- and right-eyedness in starry flounders, while genetic, *does not* confer any evolutionary advantage. Therefore, his assumption is that, as with the snail's eggs, there is some other controlling genetic trait. This unknown trait evidently does confer some advantage in the Western Pacific that it does not in the Eastern Pacific.

Another well-known example of handedness occurs in climbing vines. Those that climb by coiling, such as pole beans, must either coil to the left or to the right. Nearly all such climbing vines twine to the right, but a few, such as the honeysuckle, spiral to the left. Since Shakespeare's time at least, poets have used the opposite coiling of two vines of different species as a metaphor for a close embrace, for the opposite twisting of the two plants soon locks them together. While it is relatively easy to untangle two bean vines that have grown together, the confusion caused by a right-coiling wild morning glory (the bindweed) and a left-coiling honeysuckle that have intertwined is monumental. Shakespeare alluded to this tangle as a symbol of misguided love between Queen Titania, who was bewitched to fall in love with Bottom when he had the head of an ass, in *A Midsummer's Night Dream.* Michael Flanders and Donald Swann, however, wrote a song about the problem, called *Misalliance,* which they used in their popular comedy act of the 1950s. In the Flanders and Swann version, the honeysuckle and bindweed fall hopelessly in love, are thwarted by their parents, mocked by people who wonder which way any "offshoots" will curl, and finally succumb from the results of their love affair—stem-crossed lovers, if not star-crossed.

Some migratory birds show a form of handedness that is more akin to the kind humans have. A few Eastern warblers are spotted each year in California, far from their normal summer range. All birds found outside their usual range are called *vagrants* by bird watchers. Biologist David DeSante had an idea about the vagrant birds; perhaps they could not tell right from left.

If you were giving instructions to a warbler in Canada as to how to find New England, you might say something like "bear left about 55 degrees instead of heading due south." In fact, that is the path that most Eastern warblers take. But the vagrants, DeSante reasoned, might not know right from left. If so, it would be a matter of chance whether they headed toward New England or toward California.

This idea could be checked. If warblers are caged at the time of year they normally migrate, they tend to orient themselves in the direction toward which they would migrate if they could. DeSante was able to trap 24 vagrant blackpoll warblers and compare their orientation in a cage with that of blackpoll warblers that had made a successful migration to the East. Sure enough, the vagrants were equally likely to orient themselves to the right as to the left in the autumn, while the other warblers were adamant in turning left. (You might argue that some of the successful warblers could be vagrants who happen to be lucky, winding up on the proper coast anyway. The pool of successful warblers is very large, however, so that the chance of the sample of successful migrators including "lucky vagrants" is very small.)

Around the world, the same phenomenon exists. Most migratory birds are able to tell right from left, but a few in each population seem to get the directions backwards. It should be noted that the phenomenon is not confined to birds. Most people have no trouble as adults in automatically telling right from left, but some people have difficulty all their lives. Such people have to stop and think about which is which—a form of directional dyslexia, perhaps.

Tests by psychologists show that children confuse right and left completely at age 5, but can tell them apart about 65 percent of the time a year later. This ability reaches 92 percent by age 9, but even by age 12 there are problems in telling another person's right, although the twelve-year-olds generally have little trouble with their own. A few people, however, like me, have trouble all their lives. Often such people have trouble reading because they confuse letters such as *b* and *d* or *p* and *q*. Such a condition is called *dyslexia*.

A recent series of experiments suggests that there is, after all, a form of nonhuman handedness. In the past two years, various scientists have established what previous scientists failed to find—clear right- or left-handedness among some animals who use their paws to grasp objects and to feed themselves. It should not be surprising that the animals with such preferences turn out to be primates. It is surprising that the experimental animals that show certain types of handedness are monkeys and prosimians—our second and third cousins—not the more closely related great apes. We will come back to these discoveries after learning more about human right-left preferences.

Lopsided Humanity

Human beings are only superficially symmetrical. In most people, the heart is on the left, for example, and the appendix on the right. In a few people, however, the organs are reversed, called *inverse situs,* and the heart is on the right and the appendix on the left. Actually, the heart is just barely on the left, but it sounds louder on the left because the left ventricle makes a louder noise. In *inverse situs,* the heart and its connections are reversed, so—although still nearly in the center of the body—the right ventricle is the louder and the heart seems to be on the right.

Another human asymmetry is only present at birth. The human umbilical cord is a helix, so it must be either right-handed or left-handed. It is left-handed, like the horn of the narwhal.

The external symmetry of human beings is not perfect, either. Most actors are very aware of this, choosing to be photographed from their good side. An interesting experiment for an amateur photographer is to take a portrait and make prints two ways, one with normal right and left reversed, or flopped. With a flopped portrait, even the subject probably does not notice any strangeness when viewing his or her own reflected image, since most people are used to looking in mirrors occasionally. But if the two versions of the portrait are cut in half and skillfully put together at the midline of the face, it becomes clear that something is wrong. A perfectly symmetrical face is not quite human.

The two halves of the face also express emotion differently. A symmetrical composite of the left halves of a person's face (one flopped and the two stuck together) is seen by viewers as expressing an emotion more strongly than the same person's face expressing the same emotion in a right-face composite.

Another external asymmetry is that, for most people, the length of the right arm is slightly different from the length of the left. For some people, this is sufficiently pronounced as to affect their tailoring. Also, the right leg is a different length than the left. This phenomenon has been advanced as the reason that people who are lost in a forest, snow, or fog tend to travel in circles. One assumes that if the left leg is longer, the circles will be clockwise, while if it is the right, the circles will be counterclockwise. Furthermore, the degree of asymmetry should affect the size of the circle. It has been calculated that a person whose left leg is $1/16$ inch shorter than the right will walk in a counterclockwise circle with a radius of 720 feet if he or she is not making constant corrections. I do not know of any scientific experiments that have been made to verify this assumption, which some may doubt. The theory may fall in the same class with the Swiss mountain cattle that are said to have the pair of legs on one side shorter than the other so that they can graze more effectively on the mountain (but they always have to travel in the same direction, for if they reversed course, they would fall over).

One interesting set of asymmetries in humans, however, concerns the differences between men and women. Physically, one notable difference is that men have an asymmetry that women do not; the left testicle generally hangs lower than the right. Related to this in a vague sort of way is the fact that most men normally have a preferred side of their trousers in which to "park" their genitalia. British tailors, among the best in the world, do not overlook this detail; they ask clients who are having suits made whether they habitually "dress to the left" or "dress to the right." The answer determines how the pants will be cut. (While women are asymmetrical in terms of breast size, this is not as consistent an asymmetry.)

Recently the question of a difference in mental asymmetries between men and women has arisen. This is based upon an even more fundamental asymmetry in the human brain.

Just as the heart is somewhat on the left and the appendix generally on the right, certain mental faculties are generally found in the left hemisphere of the brain and others generally in the right. The most basic of these is that the left hemisphere controls the right side of the body and the right controls the left, but, in the 1860s, Pierre Paul Broca identified an area in the left side of brains of right-handed people that controls speech. As noted earlier, this *may* be a result of right-handed skills, such as throwing, also being located in the left side of the brain. Interestingly, control of speech is often in the right side of the brain in left-handed people.

In the 1950s, Roger Sperry found that the differences between the sides of the brain are more complex than this. The left side of the brain is the one people normally use for language and analytical thinking. The right side is the one normally used for perception of shape and form. These conclusions result from experiments with people whose brains had been surgically split as part of a treatment for epilepsy, and they were followed up by studies of people who have suffered strokes as well as experiments on people whose brains were intact. All the different approaches point to the same type of difference between the halves of the brain. One experiment, for example, used radioactive studies of the working brains of musicians compared with nonmusicians. When asked to compare tone sequences, the trained musicians processed the comparison in the analytical left-brain, while the nonmusicians processed the comparison in the pattern-recognizing right-brain. When asked to compare harmonies, however, both groups used the right-brain.

A similar study using electroencephalograms was conducted on lawyers and potters. As expected, the lawyers showed a tendency to more left-brain activity than the potters did.

Where did all this difference in the halves of the brain come from? Let's return to the recent discovery of handedness in certain monkeys and prosimians. The first question to be answered is why earlier researchers failed to take note of handedness in any primates except human beings. Apparently the answer is that the earlier researchers conducted the wrong experiments. In all cases, researchers looked at situations in which a prosimian, monkey, or ape was presented with a stationary, visible target, such as a fruit on a tree. The primate would, in that situation, reach for the target with whichever hand was convenient. It did not occur to the researchers that we human beings also tend (but only tend) to do the same thing. If I am picking peas, I may pick with my right hand and hold the container to collect the peas with my left. However, if I have a basket on the ground into which I toss the picked peas, I tend to use both hands about equally, depending on where the peas are. This old pea-picker is evidently not that far removed from any other primate when presented with a convenient, stationary target.

Researchers found that things were different when the target was moving, when the target was hidden, or when the primate was standing on two feet with nothing to cling to (for primates that usually cling, such as bushbabies and some lemurs). The ones that cling tend to reach for objects with their left hands. One might think that is because their right hands are clinging to a tree branch most of the time. This does not seem to be the case, however. Instead, they use their left hands especially when both hands are free and the primate is standing up on two feet.

Clingers and squirrel monkeys are also especially prone to use their left hands when reaching for floating or swimming things.

How the left-hand preference in some primates became a right-hand preference in others is not clear. Our fairly close relative, the gorilla, has been shown to have a left-hand preference in certain tasks that require some care. The rhesus monkey, not nearly such a close relative, shows good evidence of regular right-handedness. The major theory advanced by researchers in the field to explain all this is that originally it was the right side that was preferentially used for clinging and the left hand for grasping. When some primates began to walk regularly on four legs and stand up on two occasionally, the strong right hand—no longer needed for clinging—became the main foraging tool, while the left lost its importance. This idea would also explain why the left brain hemisphere, which controls the right side, specializes in posture. Other scientists think this does not explain very much. Melvin Goodale thinks more of a "throwing" hypothesis. In his version, primates are preadapted to throw better with their right hands, encouraging its further development.

The assumption of both of these theories is that evolution has produced the differences in the two halves of the brain. What has this to do with differences between men and women? Psychologists have found that humans vary in their abilities with spatial relationships (right-brain stuff) and language (left-brain stuff) according to sex. Women are generally better with language, while men are better with spatial relationships. Some psychologists believe that this difference is culturally determined, but there is increasing evidence of a biological basis for the difference. No one doubts that the different levels of the hormones testosterone (more in males) and estrogen (more in females) control many of the physical differences between males and females. Recent research on monkeys, rats, and songbirds has shown that in the species studied, levels of these hormones actually affect the structure of the brain during its growth. The most dramatic effect is with songbirds, for female birds given testosterone not only learn to sing the songs of the male birds, but also show development of structures in the brain that can be shown to control the song. Although no conclusive experiments have been done with humans, the indications are that physical differences account for women being more left-brained and men being more right-brained. The evidence shows that the differences are caused by different levels of the male hormone testosterone in the developing fetus. These differences are shown in a brain structure called the planum temporale, which is generally larger in women.

This difference between right-brained men and left-brained women has been used to account for the fact that more men than women tend to be left-handed. Studies have shown that (1) a higher percentage of normally intelligent children who have reading problems are left-handed, (2) a higher percentage of left-handed children have problems learning to read, and (3) of these, boys tend to have more reading problems than girls. Another study, conducted by Allen W. Gottfried and Kay Bathurst, from California State University at Fullerton, found that definite handedness—that is, choosing either the right or the left hand consistently for certain tasks—was a consistent indicator of early intelligence in girls, but not in boys. The same study also found that handedness is usually established quite early, generally before 18 months of age. One possible explanation for the difference in indications of mental precocity between boys and girls is that the study was not concerned with whether the handedness was right or left. (In fact, among the young children considered, there was only one consistent left-hander, and she was, atypically, female.) Since left-handed boys are at a disadvantage, the overall handedness in relation to intelligence might be affected. It seems more likely that many of the inconsistent males might, as they became older, become consistently left-handed.

The connection between right-brainedness, left-handedness, sex, and testosterone has also been used to explain why left-handed people are more likely to be victims of migraine headaches and a variety other neurological diseases. Testosterone not only tends to affect the handedness caused by the brain, but it may also affect the brain in other ways, causing potential neurological problems. Similarly, there is considerable evidence that certain diseases in which the immune system attacks a person's own body (instead of attacking "foreign" invaders) are closely correlated with left-handedness and with testosterone during gestation. In all these cases, evidence is building that the testosterone becomes involved by reducing the amount of growth of the left planum temporale.

Research on rats has shown that there are similar asymmetries between the brains of male and female rats. The research itself sounds like something that would win one of former Senator William Proxmire's famous "Golden Fleece" awards. Rats were strapped into a harness and given such substances as speed or cocaine. These drugs tend to cause rats to move in circles. Cocaine was the most effective in that it caused most of the rats to rotate (the rotations were measured by the harnesses). Some of the rats rotated to the left, while others rotated to the right. Female rats tended to rotate to the right, while male rats tended to rotate to the left. Also, female rats that rotated to the right rotated much more often (up to 400 rotations per hour) than female rats that rotated to the left. Similarly, male rats that rotated to the left went round in circles more often than right-rotating male rats.

The purpose of the experiments, by the way, was to find out more about the way that substances that inhibit the important neurotransmitters (such as dopamine) in the brain affect the brain, with specific implications for understanding drug addiction. The researchers also concluded that the amount of dopamine suppressed in the two sides of the brain was different, and that female rats (at least) exhibited a different asymmetry (in general—*not* always) than male rats.

Earlier studies had shown that rats not given drugs exhibited much the same asymmetries, especially with regard to the directions in which they normally turned. This direction is set at birth—and, astonishingly, can be determined by which way the newborn's tail bends. As in the cocaine study, females mostly prefer to turn to the right, while males are far more variable. Also, as in the cocaine study, the amount of dopamine in the two sides of the brain seemed to be directly involved in asymmetric behavior.

Not everyone agrees with either the evolutionary theories of development of the two different functions for the halves of the brain or with the theory that testosterone causes the differences. Marian Annett of England, for one, thinks that the difference is present only in humans (and hence not an adaptation enhanced by evolution) and that it is a direct result of a genetic characteristic (not a developmental one, such as testosterone in the womb). She points out that many people are right-handed for some functions, but left-handed for others. She thinks that most people inherit a gene for left-brainedness, but not the 90 percent that are right-handed. If only 60 percent inherit the gene and the rest split fifty-fifty, it would almost explain the observed proportions. This theory seems to fail to account for differences between men and women, unless she assumes that the gene involved is somehow more likely to be found in the Y (male-determining) chromosome.

Looking at it another way, the Y chromosome is smaller than the X chromosome, of which two are needed to produce a female mammal. Suppose that either a left- or right-brain gene is located on the part of the X chromosome that has no matching genes on the Y. Also assume that the left-brain gene is recessive while the right-brain gene is dominant—you need two left-brain genes to be left-brained, but only one right-brain gene to be right-brained. Chance alone would suggest that about 50 percent of the men would be left-brained, since they have no matching gene. On the other hand, with a 50 percent chance in each X chromosome of getting the dominant right-brain gene, a woman would stand a 25 percent chance of being left-brained. Since the 90–10 split is not exact for all functions, this might be considered close enough.

While such a scenario would tend to account for observed differences, it is far from proved. Perhaps when the human genes are completely identified, we will learn the truth.

All of this is another way of saying that there is more to line symmetry than that the thing on one side looks just like the thing on the other.

How *Does* It Do It

When Alice stepped into the looking-glass world, everything was reversed from left to right. It would not have been much of a story if everything were reversed from top to bottom, would it? We are so used to the knowledge that mirrors reverse from right to left that we would certainly not expect top-to-bottom reversal. Yet, when you start to think about why mirrors reverse from right to left and not from top to bottom, your thinking is apt to get a bit topsy-turvy itself.

In fact, the question of why a mirror reverses from left to right and not from top to bottom has engaged various philosophers who have put forth conflicting theories to account for the reversal. Their several explanations were published in such journals as *American Philosophical Quarterly, Journal of Philosophy,* and *Philosophical Review.* All this suggests that it is not an easy question to answer.

The reversal *is not* a physical property of the mirror. All flat surfaces that reflect light show the same reversal, whether they are made from glass, silver, or water. Also, if you rotate a small mirror a quarter-turn in either direction, the reversal will still be from left to right, not from top to bottom. Alternatively, you can lie on your side in front of a mirror; the result will be a reversal of *your* left and right, although this time the reversal will actually be from top to bottom. This is an important clue to what is happening.

In fact, the mirror does not really reverse your left and right when you are standing, does it? If you raise your right arm, the image on the right side of the mirror will show an arm raised. Granted, if the image were an actual person, that would be the person's left arm, but it is your right arm and it is still on the right side of the mirror. If you did not know whether you were looking at a person directly or in a mirror, you could not tell whether the person was left-handed or right-handed.

The ability of a mirror to reverse left and right is shown more clearly with ordinary writing, since the mirror image is almost unreadable without practice. Also, writing in English always proceeds from left to right, but the image is from right to left. Even a word with all symmetrical letters, such as MOW, becomes WOM. If you write an ordinary sentence, such as "WHY IT'S A LOOKING-GLASS BOOK, OF COURSE!" and reflect it in a mirror, not only will it show reversed letters in the order "!ESRUOC FO ,KOOB SSALG-GNIKOOL A S'TI YHW" but the letters themselves will be backward. In such a case it would be easy to tell whether you were looking at the original or the image, unlike the case of looking at a person.

If the mirror reversed from top to bottom, the letters would be reversed, but the order would be the same as in the original. You can determine this by turning the paper upside down and looking at the reflection. If the letters in a sentence are all symmetrical about a horizontal line, the reflected upside-down sentence will read normally. Examples of such sentences are not necessarily easy to come by, but there are a few. For example, if you wanted to describe the profession of a bootlegger in the 1920s, you might say

HE BOXED CHOICE HOOCH.

HE BOXED CHOICE HOOCH.

Or, in describing your relation to your assistant when you were OSS spies during World War II, you might remark

HE DECODED, I DECIDED.

HE DECODED, I DECIDED.

Yet, if you turn "WHY IT'S A LOOKING-GLASS BOOK, OF COURSE!" upside down, the S's, for example, will still appear reversed, unlike the C in "CHOICE," which no longer appears reversed when it is reflected upside down. In fact, the C in CHOICE appears reversed from left to right if you view it directly upside down, but S appears normal when seen directly but upside down. Thus, the upside down C, when reversed left to right by a mirror, is returned to its original appearance, but the upside down S looks as "backward" as it did when it was rightside up.

What is needed for clearer understanding of the mirror phenomenon is a model, something that is easier to understand than a mirror but which behaves in the same peculiar way. Fortunately, such a model is easily found. Write the sentences WHY IT'S A LOOKING-GLASS BOOK, OF COURSE! and HE BOXED CHOICE HOOCH on a sheet of thin paper. Now look at the sentences through the paper with the aid of a strong light. They will be reversed exactly as in a mirror. If you turn the paper upside down, HE BOXED CHOICE HOOCH will read correctly through the back of the paper. Viewing the letters from the back is a model of seeing them in a mirror.

Thus, it would appear that the mirror reverses not left and right, but front and back. This conclusion, which in some sense is certainly correct, is a bit unsatisfying. If a mirror reverses front and back, it is easy to see why it does not reverse top and bottom; on the other hand, when you look in a mirror, you do not see the back of your head.

Part of the problem is that writing is two-dimensional, but people are three-dimensional. Standing in front of a mirror with a sentence such as "WHY IT'S A LOOKING-GLASS BOOK, OF COURSE!" written on a piece of clear plastic shows that the image you see through the back of the plastic is exactly the same as the image in the mirror. If the words are normal from the side of the plastic you are on, and the plastic is between you and the mirror, the words will appear normal in the mirror. On the other hand, if you put yourself between the plastic and the mirror, one sentence will appear normal and the other will not. If the sentence on the plastic is normal, its reflection will be mirror-reversed, and vice-versa.

An observer watching you looking at yourself in a mirror sees a similar phenomenon. If the observer is behind you, he will see your right hand on the observer's right with the image of your right hand also on the observer's right. If you turn around to face the observer, your right hand will be on the observer's left and also the image of your right hand will be on the observer's left. If the observer is between you and the mirror, however, when the observer faces you as you look at the mirror, your right hand will be on the observer's left, and when the observer faces the mirror, your right hand will be on the observer's right. Even for three-dimensional objects, the mirror is reversing from front to back, not from right to left. We perceive the reversal as from right to left when something with a known right-left asymmetry is seen in the mirror *simply because we don't make the psychological adjustment that we are seeing the object from a different point of view.*

Writing in the mirror appears reversed because you are seeing it as if you were reading it though the back of the paper. This is not to say that the reversal is purely psychological. From your point of view, the writing is reversed. On the other hand, from your image's point of view, the writing is perfectly normal. It is not reversed from top to bottom.

Most people, including me, find this way of thinking about a mirror difficult to do. It is much easier to think of a mirror as reversing left and right than it is to think of seeing what your image sees. Nevertheless, if you can make this adjustment, you can understand why a mirror appears to reverse right and left and does not appear to reverse top and bottom.

Lewis Carroll seems to have overlooked this point in *Through the Looking Glass;* he does not have Alice look back through the looking glass at the room in the real world. If she had, what would she have seen? The real world would appear to be reversed, while the looking-glass world would have been normal. Probably the mathematician Reverend Dodgson knew this, but the author Lewis Carroll chose to ignore it in the interest of making a better story.

Contrariwise

One of the first things Alice looks at is a clock, but the looking-glass clock has an old-man's face instead of a clock face. Carroll missed a bet here. A clock viewed in a mirror is an excellent example of apparent reversal. The hands appear to be moving counterclockwise because the point of view is like that of seeing a transparent clock from its back. Clockwise and counterclockwise are thus even more clearly differentiated than right and left, but they behave in the same way in a mirror; they appear to be reversed. If you saw the face of the clock from the point of view of someone inside the mirror (that is, from the point of view of your image) the clock hands would move in the normal clockwise fashion.

Clockwise seems so clearly the positive direction to most people that counterclockwise patterns are seldom encountered. Most screws or nuts, for example, turn clockwise, although situations involving screws or nuts on a rotating body sometimes necessitate making counterclockwise threads to keep the screw or nut from working loose. Does the wheel on a car turn clockwise or counterclockwise when the car is moving forward? It depends on your point of view; if you look at the car from the car's right, the wheels turn clockwise, but seen from the car's left, the wheels are turning counterclockwise. In some cars, nuts that turn clockwise are used to hold the wheel to the axle on the car's left, but counterclockwise nuts are used on the car's right. Screws and nuts that turn clockwise are called right-handed, while screws and nuts that turn counterclockwise are called left-handed.

Are horse races run clockwise or counterclockwise? Oddly enough, the answer to this depends on whether you live in the United States or in England. Just as the English reverse our concept of which side of the street a car should be driving on (and consequently, English-made cars, unless specially made for export, are mirror images of American cars), the English run races clockwise, but Americans run their races counterclockwise. Consequently, 40,000 American improvers of the breed were startled when they came to Belmont in New York City for the first time on May 4, 1905. Perhaps in an effort to be "tony" by being British, the races were run in the wrong direction—clockwise. It was not until 1921 that Belmont began to run races in true-blue American fashion. This seems to be a reversal of the general rule for clockwise being positive and counterclockwise being negative, but there are other such reversals.

One of them is peculiarly mathematical. As anyone who has ever taught trigonometry to high-school or college students knows, one of the hurdles to overcome is that mathematicians view counterclockwise as positive while clockwise is negative. When an angle is defined as an amount of turning, as it is in trigonometry, the angle has a positive measure only if the turning is in the counterclockwise direction. Students find this approach uncomfortable at first but soon become used to it.

Acid with a Twist, Please

The side a flounder turns on, the screwiness of a narwhal's horn, or the direction a plant twines may seem like trivial examples of the fundamental asymmetries of life. But there is one asymmetry that is even more basic than that, even more basic than any differences between masculine and feminine brains. The very molecules of life are asymmetrical. To detail this asymmetry, however, requires a long story.

Ancient people were quite familiar with transparent substances, since quartz, a common mineral, is transparent. Whether or not they were also familiar with one peculiar property of another transparent mineral that we are about to discuss is not known (although I think they could hardly have missed it). This property did not reach the attention of science, however, until 1669, when Erasmus Bartholin (1625–1698) reported that Iceland spar, a transparent crystal occurring in limestone, shows double images. If, as shown in Figure 2–1, a crystal of Iceland spar is placed over two intersecting lines, each of the lines will be doubled when viewed through the crystal. This property is called *double refraction.*

Ordinary refraction occurs when light is bent because of the change of speed that occurs when light leaves one transparent substance and enters another. Double refraction could only be explained in a single crystal if light came in two different forms, one of which was refracted at a different angle from the other. Among the people who commented on the phenomenon was Sir Isaac Newton, who thought that perhaps light particles came in

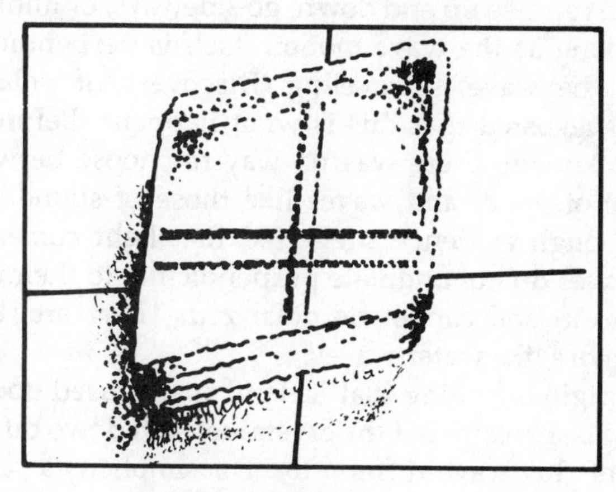

FIGURE 2–1
Iceland spar exhibits the phenomenon of double refraction.

two forms, like the north and south poles of a magnet. Newton did not pursue this idea, however.

In 1808, Etienne-Louis Malus (1775–1812) discovered that sunlight that had been reflected from a distant window was not doubly refracted by Iceland spar. He knew of Newton's idea, so he explained his discovery by saying that only one pole of light had been reflected from the window. Since Iceland spar split regular light into two poles, but this light did not split when seen through the crystal of Iceland spar, he concluded it had only one polar form in it. He called this light *polarized* light, a name we still use today, although the idea that light has two poles has been completely abandoned.

Polarized light could be better explained in terms of waves. Suppose that light is a wave similar to a wave in water or to a sound wave. Waves in water just go up and down, because of gravity (that is, the wave rises because of some original outside force, such as the wind; then falls because of gravitational force; then rises because of the forces causing a liquid to seek a level—but rises above the level, and falls back because of gravity). Sound waves go back and forth, not up and down. Light waves are different from either sound or water waves. Light waves are not necessarily constrained to one plane as water waves are. A light wave can rise up and down, go sideways, or undulate at any angle so long as the wave motion itself is perpendicular to the direction the wave is traveling. Discovery of polarized light strongly suggested that this is what happens. Before polarized light was known, there was no way to choose between waves like those of water and waves like those of sound to describe light, although evidence suggested that light comes in waves. Sound waves do not undulate perpendicular to their motion, but parallel to it, and cannot be polarized. Therefore, light waves must be more like water waves.

The original finding that Iceland spar caused double refraction could be explained if the crystal refracted two different components of this wave differently. For simplicity's sake, assume that the two components are up-down and right-left. The up-down component could be refracted at a different angle in Iceland spar than the right-left component. Similarly, the window

Malus saw through Iceland spar might reflect just one component of light, so it would produce just one image. That is just what happens, as we know today.

Malus discovered polarization in 1808, and some investigators began to look for other ways to produce polarized light. In 1815 Jean-Baptiste Biot (1774–1862) found some substances that would twist a beam of polarized light passing through them. Among the substances were solutions of various kinds. A few years later, in 1828, clever devices were constructed to measure the amount of twisting in polarized light, instead of merely detecting its presence. With careful measurement, the ability of a substance to twist polarized light could become one of the characteristics of the substance, along with melting point, color, conductivity, solubility, and so forth.

Biot soon uncovered a major mystery. Two different batches of what seemed on all other grounds to be the same substance were different with respect to polarization. One of them, common tartaric acid, twisted polarized light in solution to the right. The other, now known as racemic acid, did not twist it at all. By then, chemistry had advanced to the point that it was possible to determine the elements that combined to form these two apparently different substances (when twisting of polarized light was used as the measurement). The elements were the same and in the same amounts.

At the time, it was the outstanding mystery in chemistry. A brilliant and dedicated young chemist, just entering the profession, sought to solve the mystery.

The scientist was Louis Pasteur, better known to the general public today as a medical researcher. In fact, it was his study of racemic acid that pointed his work toward the understanding of microorganisms and the development of vaccines for which he is best known. In 1844, while still a student, he began working on the racemic-acid problem as one of the two pieces of research he was to submit for his doctor's degree. At the age of 25, he had solved it.

Not only were the components of racemic acid and tartaric acid the same, but also all the other properties associated with a substance—except for the twist tartaric acid gave to polarized light. But, by close examination, Pasteur found one other difference, a difference that other scientists had missed. Although it was believed that the crystals of the two substances were identical, Pasteur noticed that the crystals of racemic acid came in two enantiomorphic forms—one just like the crystals of tartaric acid and the other with small facets on the opposite side, a mirror image of tartaric acid. Pasteur carefully separated the two enantiomorphs by manually picking the crystals apart under a microscope, producing small sets each including only one of the enantiomorphic forms. Then he dissolved each set of crystals and observed what happened when he passed polarized light through the two solutions.

The solution of the crystals that had the same handedness as tartaric acid behaved just like tartaric acid, twisting polarized light to the right. The other solution did something that had not been seen before in either tartaric or racemic acid. It twisted a beam of polarized light to the left. Pasteur had discovered a third version of the chemical.

He also had partly solved the mystery: Since racemic acid was composed of roughly equal amounts of an acid that twisted light to the right and an acid that twisted light to the left, racemic acid itself (now seen as the combination) did not twist light at all.

Pasteur guessed at the remainder of the solution to the mystery. His guess was exactly right, as was shown by other chemists 30 years later; tartaric acid and other chemicals that twist polarized light are asymmetric at the molecular level.

Two other chemists, Jacobus Hendricus van't Hoff and Joseph Achille Le Bel, determined in 1874 how molecular structure could produce two different enantiomorphic versions of the same chemical. They suggested that the four bonds of a carbon atom—that is, the places where other atoms attach to form a molecule—are in two different planes. If two bonds are in each plane, you have a situation like that of a two-dimensional coordinate system: There are four different components. In the coordinate system we think of these as $+x$, $-x$, $+y$, and $-y$. The four components can be arranged in two different mirror-image forms.

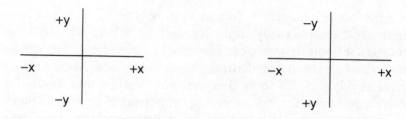

These are the two forms that rotate light either to the left or right. In general, the kinds of chemicals that do this will involve carbon atoms to which four different atoms or groups of atoms (called *radicals*) are attached, one to each of the four bonds. If two or more of the atoms or radicals attached to the carbon atom are the same, then the molecule can exist in a single form. Such a molecule can be rotated into all its possible configurations. If none of the four atoms or radicals attached to the carbon atom are alike, however, there are two possible forms; and no amount of rotation will turn one form into its mirror image (Figure 2–2). Thus, most enantiomorphic molecules are complex molecules based on carbon.

But for a long time it was impossible to tell whether the compounds that were labeled as D- (for *dextro-*, or right) or

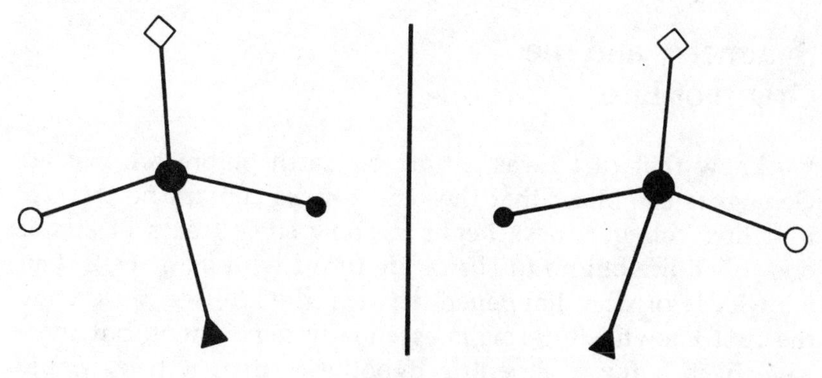

FIGURE 2–2
Mirror images of carbon atoms with four different radicals. No amount of rotation will convert one into the other.

the compounds labeled L- (for *levo-*, or left) really were the ones that were respectively right-handed or left-handed at the molecular level. It was clear how they rotated light, but it was much less clear what structural properties caused them to affect light in this way. To solve this problem, Emil Fischer devised a system in which the light-rotating properties of chemicals were related to the properties of the two enantiomorphs of the chemical glyceraldehyde. This method assigned some left-handed light-rotating chemicals to a right-handed series and vice versa; for instance, tartaric acid, although it twisted light to the right, was found to have the same handedness as the left-handed form of glyceraldehyde. Thus, the regular form of tartartic acid, produced in the fermentation of grapes into wine, was labeled as L-tartaric acid. Finally, in 1955 it was shown that Fischer had guessed right and that the classification of left-handed and right-handed chemicals he had developed made sense at the molecular level, even though it was inconsistent at the level at which people measure the turning of light.

The idea that there are left-handed and right-handed molecules has been useful in a number of ways. But one of the most interesting of the applications has been in trying to solve the mystery of the origin of life.

Symmetry and the Origin of Life

We know that there was a time on Earth before life existed. Geologists can show that the oldest rocks contain no trace of past life. Younger rocks begin to show some traces of simple one-celled life but no fossils of life forms with many cells. Our knowledge of what happened between the life-free period and the first known life forms is essentially nonexistent, but there have been a few compelling hypotheses that, if true, would neatly fill the gap. Here are some of the more prominent of these speculations:

1. *Life originated somewhere else in the universe.* Cosmic dust or meteorites carried that life form to Earth in the form of spores, specially adapted to withstand extremes of heat or cold and long periods without food or water. Upon encountering the Earth's oceans, the spores turned into cells that could take energy from the environment and begin to reproduce.

This hypothesis merely shifts the problem of the origin of life from Earth to some other locale.

2. *Life originated when naturally formed fats called phospholipids formed tiny, persistent bubbles in the sea.* Because one side of these large fat molecules is water-loving (hydrophilic) and the other side is water-hating (hydrophobic), such tiny bubbles always form in water to which phospholipids have been added. Water and other fluids inside the bubble are effectively separated from liquids outside the bubble. Gradually, in some bubbles, because their molecular structure permitted some molecules or ions to enter the bubble but not to leave, certain chemicals began to accumulate in the bubble. These bubbles grew in size, also; although when they reached a certain size, they would break in two. When that happened, the accumulated chemical mixture passed into the two daughter bubbles. In time, bubbles that persisted longer or that grew faster tended to develop. The various mechanisms of life appeared within the bubbles as ways of maintaining them and of encouraging their growth and division.

The bubbles were *not* especially trying to perpetuate themselves. The only mechanism at work was "survival of the fittest," which has often been called a tautology since "the fittest" are defined as those who survive. Tautology or not, it is clear that if some bubbles last longer and produce more descendants than others, eventually the bubble population will consist entirely or largely of those types of bubbles. The same mechanism is invoked in the remaining two hypotheses. It is generally familiar as *natural selection,* the mechanism behind Darwinian theories of evolution.

3. *In the early atmosphere, natural sources of energy, such as lightning or ultraviolet light, combined the simple atmospheric*

chemicals into more complex forms. The more complex forms included many of the molecules that are now essential for life, such as amino acids or even such polymers as ribonucleic acid (RNA). These organic molecules dissolved in the sea. Over long periods of time, more and more of the organic molecules were dissolved, making the primitive ocean a kind of weak soup. Nothing ate this soup, however, as there were no living things. That is, there were no living things until one of the large molecules developed by chance into a template that, with suitable other molecules plucked from the soup, formed copies of itself. Such a process is exponential. One molecule became two, two became four, and so forth. As long as there were suitable molecules to use for "food" and no competition, the process quickly created the maximum number of these molecules possible in the soup. Generally it is thought that the molecule that achieved this stage of life's development was a form of RNA.

RNA is not very stable in such an environment. Its gradual breakdown and reassembly allows for minor changes in its structure that tend to be "inherited" (i.e., to reappear when new RNA is formed). Some of these changes made one line of RNA more "successful" than another, since the more successful RNA had one or more of these characteristics: forming faster; breaking down some common but useless molecule to get at a useful submolecule; catalyzing a reaction between two common useless molecules to produce a useful one; developing structures that tend to make it more stable; developing a way to use a photon from sunlight to raise the energy level of a substance; and so forth. Such changes gradually accumulated in the lines of successful RNA's until something that we might recognize as primitive life developed.

This scenario has the interesting consequence that life of this type could not evolve today. For, if life existed at all, the soup of organic chemicals would quickly be consumed. Only if there were no organic life could such a soup be created.

4. *Defects in the microcrystalline structure of some clays can reproduce the same structure over and over again.* If there are two slightly different versions of such a structure, the one that will persist better in a difficult environment will replace the less persistent one throughout that environment—natural selection, again.

Such defects have the essential characteristics of genes. They can make copies of themselves, or replicate (*replicating* is generally used for inanimate forms, while *reproducing* is saved for animate ones). The copies can be imperfect, so that evolution can take place. As a result of such evolution, these replicating defects gradually develop protective alliances with specific secondary clays, forming a tiny body or cell. When a gene or collection of genes directs the formation of a living thing, the organism is called a *phenotype;* by extension, this "body" of a clay defect is also called a phenotype.

The defect may find it particularly useful, for faster propagation or for stabilization, to incorporate organic molecules into its phenotype. Perhaps one of these is RNA, incorporated for its structural properties. RNA is a long molecule of indeterminate length that has a negatively charged backbone. Such a backbone would tend to stick to the edges of the clay particles. As the phenotype evolved, the RNA became a secondary method of preserving genetic information. But RNA is better at preserving information than the original defect was, so RNA took over as the gene in the cell. The original defect gradually withered away.

The result was life. That is—to summarize hypothesis 4 in another way—life as we know it today was built on a template formed by inorganic substances that had grown to be complex as a result of evolutionary forces. The main difference between this hypothesis and hypotheses 2 and 3 is that in those hypotheses you expect to find processes similar to today's cell processes at very early times and one expects the same kinds of molecules to be involved from the beginning. In hypotheses 4, you expect that prelife evolution will proceed according to different processes and involve different kinds of molecules.

For one of these hypotheses to be accepted as the theory of the origin of life, it must explain the characteristics of life that we see today. It is first necessary is to establish one or more characteristics of life that distinguish it from the inanimate world.

Interestingly, Pasteur was one of the many people who speculated about the difference between life and nonlife. Most of such speculations have involved *vitalists*, people who believe that there is some fundamental difference between living and nonliving things, as opposed to *mechanists*, people who think that the physical processes that occur in nonliving entities can also explain all the mechanisms of life. Strictly speaking, Pasteur was a vitalist, although not much like the garden-variety. Typical vitalists believe in some mysterious, unknown, and often unknowable, quality that separates life from nonlife. Pasteur, however, thought he knew exactly what separated the quick from the dead. Although he could not prove it, he thought the vital principle was *lack of symmetry*.

Recall that Pasteur's work on polarized light involved tartaric acid produced by grapes and racemic acid produced by chemical reactions. The tartaric acid rotated light to the right, while the racemic acid was not optically active. Pasteur was able to produce two optically active forms of tartaric acid, however, by physically separating two forms of crystals found in racemic acid. In other words, a living thing—a grape plant or Pasteur himself—was needed to produce the asymmetry that caused optical activity. Since all of the optically active materials then known were the products of life in some way, Pasteur speculated that asymmetrical molecules were the secret of life.

In 1857, Pasteur found a remarkable piece of circumstantial evidence for his theories about asymmetry and life, although he did not know enough at that time to recognize it as such. It was known that a salt of tartaric acid might, if contaminated with protein (or "albumin," as substances with high protein were then called), ferment. At that time it was far from clear that fermentation was a living process, as we know it to be today. Nevertheless, Pasteur deliberately contaminated an optically inactive salt of racemic acid to see if it would also ferment. It did ferment, and as the fermentation slowly changed the material, it also became optically active. Unlike the salt of tartaric acid, however, which twisted light to the right, the fermented salt of racemic acid twisted polarized light to the left.

Even though Pasteur did not know at that time what caused fermentation, he was so convinced that only life could produce asymmetry, that he began to suspect that fermentation had to be a living process. This idea, coupled with some fortuitous requests from French industry, was to lead Pasteur to his great discoveries in biology and medicine.

Later researchers were able to track down in detail how the asymmetry Pasteur noted was formed in living creatures. Life depends on complex chemicals called proteins as the active players in the processes that sustain it. Proteins are made from strings of subunits, just the way that written books are made from letters, spaces, and punctuation. Just as there can be an uncountable number of different books formed from 26 letters, a space, and a few punctuation marks, there is also an uncountable variety of proteins that could be formed from 20-odd subunits.

The 20-odd subunits are fairly simple compounds called *amino acids.* They are based on carbon and have enough different components—nitrogen, hydrogen, oxygen, and various radicals formed from these elements—that they generally can exist in two optically active forms (the amino acid glycine is an exception). As with tartaric acid, however, only one of these two forms occurs naturally in living creatures—the L-form. As a result, the proteins, which are chains of amino acids, form as helixes—connecting a long string of L-amino acids gives the string a definite twist. The resulting helix is always right-handed.

This is convenient since the molecule that directs the connecting process is also a right-handed helix. RNA is also a long molecule formed from many subunits. In the case of RNA, there are only four different subunits, called *nucleotides.* Each nucleotide consists of a base that is either a purine or a pyrimidine, a sugar, and phosphoric acid. The four different nucleotides are identified by their bases—cytosine, uracil, adenine, and guanine. All four nucleotides are left-handed, which, as in the case of proteins, produces an RNA molecule that is a right-handed helix.

Again this is convenient, since the molecule that directs the formation of RNA, a molecule known as deoxyribonucleic acid, or DNA, is also a right-handed helix—or rather, a pair of right-handed helixes linked together. DNA is also formed from four nucleotides. It is different from RNA in the sugar component of each of the nucleotides and in that a different base—thymine—occurs in place of uracil. One result of these two changes is that DNA tends to be much more stable than RNA. Not only is DNA a right-handed double helix, but also it is long enough to coil a second time, into a right-handed superhelix.

By now the story of how DNA carries and transmits the genetic message has become familiar to the point where people are apt to accept its mystery without examination. For the following brief account, try to see the process with new eyes and you will see how amazing it actually is.

DNA consists of not one, but two, right-handed helixes bound together. The configuration of the four bases is such that an adenine on one helix must pair with a thymine on the other, while a guanine always pairs with a cytosine. As a result, the two helixes can separate and each half can "direct" the formation of a new DNA molecule. This is the mechanism by which the code can be transmitted from cell to cell.

But the same device can be used to form an RNA molecule. If the DNA "unzips" for part of the way, a much shorter molecule of RNA can (and generally does) form on one of the helixes, always the *same* one, in fact. The RNA molecule, called messenger RNA, then carries the same sequence of bases (with the exception of uracil replacing thymine) as the other strand of the DNA helix. This RNA then moves through the cell. During the next stage, slightly different processes take place in bacteria and in other kinds of living creatures. In organisms other than bacteria, the RNA is tailored as it moves toward a rendezvous with a protein factory site. Long, apparently useless, parts of the molecule are deleted and discarded. Short fragments of RNA, called transfer RNA, attach themselves to the remaining strand. Each of these short fragments attaches by the usual rule of pairing, but these short fragments each attach by pairing with exactly three bases on the longer RNA strand. The other end of each of the short fragments grabs a passing amino acid. The result is that the short fragments move a specific sequence of amino acids into just the right position so that bonds form between the acids and a protein is created.

Note that all of this activity is accomplished largely as a result of the geometric, three-dimensional shapes involved. Furthermore, the final product, the protein, often acts the way it does in the cell as a result of its overall shape as well. Thus, the fact that all these molecules are based on right-handed helixes is essential for the system to work properly.

On the other hand, it is clear that the whole system would also work properly if *everything* were reversed. DNA could be made of the right-handed versions of the bases. It would then twist into a left-handed double helix. Given the right-handed bases and right-handed amino acids, left-handed RNA and left-handed proteins would result. In fact, it is what Alice should have encountered in the looking-glass world.

Alice was suspicious of this at first. Just before she passed through the mirror, she worried about milk for Kitty in the looking-glass world. "Perhaps looking-glass milk isn't good to drink . . ." Much later, Alice tries some bread, but finds it "very dry." No doubt it was, if made from left-handed proteins. Normal living things cannot deal with molecules of the wrong handedness. This was observed by Pasteur in his fermentation experiment with a salt of racemic acid. We now know, as Pasteur himself later discovered, that fermentation is caused by the feeding of microorganisms, in this case bacteria. The bacteria consumed the right-handed form of tartaric acid, the naturally occurring type, only. Therefore, the remaining molecules in the salt of racemic acid, indigestible to the bacteria, were left, and the compound became optically active.

A few years ago, a scientist announced that he was going to produce the perfect artificial sweetener based on this same idea. Sucrose, or common table sugar, is optically active, since it exists in both L- and D-forms. The form we normally consume is produced by sugar cane or by beets and is all the D-form. The plan was to produce large quantities of the L-form of sucrose as a diet aid. Although L-sucrose tastes as sweet as D-sucrose, the body does not know how to metabolize L-sucrose. Exactly why L-sucrose has not made it to the family sugar-bowl is not clear. Perhaps looking-glass sugar isn't good to eat.

Given the overwhelming handedness of life on Earth, it would appear that some explanation of this ought to be available. Certainly, of the hypotheses about the origin of life, only one has much of a chance of explaining the pattern. All ordinary chemical processes produce equal numbers of right-handed and left-handed molecules. Thus, the origin of life in fat bubbles, a chemical soup, or a clay matrix would have resulted in mixed molecules right from the start. If life started from a single spore from outer space, however, it is possibly just the luck of the draw that the original spore was right-handed.

In 1898, before scientists had any evidence that life could have originated by chemical means, the chemist Francis Robert Japp of the University of Aberdeen, Scotland, used life's asymmetry as the principal argument for supernatural creation by a sentient god. "Only the living organism with its asymmetric tissues, or the asymmetric products of the living organism, or the living intelligence with its conceptions of asymmetry, can produce the result. . . . No fortuitous concourse of atoms, even with all eternity for them to clash and combine in, could compass this feat of the formation of the first optically active organic compound." Exactly why a living intelligence is needed to contemplate asymmetry escapes me, but the rest of the argument has merit. Chance can—and does—produce an optically active *molecule.* But chance cannot produce a large number of optically active molecules that have the same handedness.

Is a large number needed? Although a virus can reproduce from just a single molecule of DNA or RNA, a virus requires the mechanisms of the cell it infects to accomplish this. It seems unlikely that viruses are much like the first forms of life. Instead, they appear to be escaped fragments of DNA or RNA that originated within the cell and that retained the ability to make a protein. Few if any scientists think that there was some magic moment when a molecule of, say, RNA suddenly became alive and propagated its kind across the soupy seas. Instead of the magic moment, it is far more likely that many small advances were made along the path from inorganic matter to organic life.

Even if that moment did exist, it would fail to explain why today's RNA is all right-handed. That first RNA would have to have been built up from whatever nucleotides were around. Since life did not exist before that mythical moment, there would have been equal numbers of left-handed and right-handed nucleotides. An RNA molecule sufficiently long to start the great chain of being would have many, many nucleotides. How could they all be of the same handedness unless there was some mechanism to guarantee that none of the wrong sort slipped in? Recall that RNA's helical form is the result of its being formed *entirely* of left-handed nucleotides.

Scientists have offered some not-very-convincing ideas that might account for life's handedness. One of the first to do so was Pasteur. Two of his ideas were that the Earth's magnetism caused the asymmetry and that the apparent passage of the Sun from east to west might be the cause. As a good experimenter, he tested these ideas. Neither growing organisms in a reversed magnetic field nor growing them with the Sun apparently traveling from west to east produced any effect. (The Sun's apparent path was changed with mirrors.) Other ideas—not from Pasteur—included the notion that life originated in one hemisphere and was given its handedness by the Coriolis effect; and that a hypothetical separation of the Moon from the Earth had produced an asymmetrical jerk that affected molecular formation.

The nearest to a plausible idea to explain life's right handedness is based on the interaction of optically active molecules with light. Since light behaves differently for right-handed and left-handed molecules, it is not surprising that polarized light can in turn affect the handedness of molecules being formed. This effect also requires a magnetic field, so perhaps Pasteur's ideas about magnetism and light were not so far wrong after all.

Reflected light is polarized. The Earth has a magnetic field. Thus, light reflected from the ocean may have influenced the handedness of molecules in the chemical soup. Although this is the most plausible idea put forward, it is still far-fetched, especially since light reflected from the ocean is only weakly polarized. Also, the theories that polarized light is supposed to illuminate involve molecules forming *in* the oceans, where such light that is present is direct, not reflected.

Another thought has been that life originally came in two forms—right-handed and left-handed. Somehow, perhaps in another magic moment, the right-handed form gained some evolutionary advantage that was so devastating that it completely eliminated the other. Not only is this poor evolutionary theory (examples of complete removal of a whole class of creatures and all their descendants are rare), but it overlooks the presumption that amino acids and nucleotides must have existed before either proteins or nucleic acids. Since no one thinks that these were living chemicals, they would have presumably existed in equal numbers of right- and left-handed forms. Proteins or nucleic acids formed from such building blocks would twist higgledy-piggledy instead of forming neat spirals. Thus, a mechanism is needed that works earlier in life's story than the creation of the first molecule of RNA.

Where does that leave us? Not far from where Japp was in 1898. Physical theories that account for many aspects of life fail completely to explain right-handedness at the level of proteins and nucleic acids. The fat-bubble hypothesis does not address the matter in any way. The primordial-soup hypothesis cannot provide—so far—any reason why handedness would arise, although it can suggest ways to preserve handedness after it has arisen. The clay-defect hypothesis has a long way to go to explain how handedness came from a prelife based on inorganic molecules, since inorganic molecules do not exhibit handedness themselves.

Perhaps biology will someday solve the general problems of handedness, ranging from why the human is generally right-handed to why the flounder lies on a particular side in the mud. Or perhaps the coil of DNA and the twist of the narwhal's horn will be mysteries that will continue to elude our understanding.

3

A Transformation of Space

Symmetry is the property of geometrical figures to repeat their parts, or more precisely, their property of coinciding with their original position when in different positions.

E. S. Fedorov

The most curious part of the thing was, that the trees and the other things around them never changed their places at all: however fast they went, they never seemed to pass anything. "I wonder if all the things move along with us?" thought poor puzzled Alice.

Lewis Carroll

A number exists as a single entity. That is, you are likely to accept the fact that there is only one number 2 around, even if you are looking at a sheet of paper that has "2" written in several places. This curious dichotomy is resolved by the convention that the number 2 is some abstract Platonic idea hanging about in space, while the actual numeral "2" consists of molecules of a black substance that has been attached to a sheet of paper in a particular pattern.

In geometry, the same Platonic idea seems to hold some of the time and not other times. If I tell you that there is only one triangle that has a unique set of three sides, you might or might not agree that I was telling you the truth. If you have studied geometry before or if you have experimented with drawing triangles that have a given set of sides, you will accept that there is only one triangle in the same Platonic sense that there is only one 2. On the other hand, when you start to solve problems or do proofs in geometry, you are confronted with a multiplicity of triangles that must be proved congruent to get the job done. In what sense, then, are two of these *different* triangles really *the same* triangle. You can draw several different versions of "the same" triangle on a sheet of paper. Are there a whole lot of triangles that have a set of sides that are the same length, or is there really only one?

Similarly, to say that a point is an "exact location in space," as many textbooks do, is susceptible to serious questions about what that means. As with "the same," it is not clear what is meant by "exact location." For example, if you have indicated a point on your paper, and someone later moves the paper, is it then a different point? In any case, isn't the Earth whirling around through space at a great speed? The more you think about these matters, the more confusing they become.

Sometimes the only way to sort out complex ideas is to find a totally different way of looking at the subject. For geometry, the first person to find a comprehensive and satisfactory way to deal with these issues on a practical as well as a theoretical basis was a mathematician named Felix Klein (1849–1925). In 1872, he proposed what has come to be known as the *Erlanger Programm,* since it was first announced in Klein's inaugural presentation after he had become a professor at the University of Erlangen. In the *Erlanger Programm,* Klein defined a geometry as the set of properties that are invariant under a particular group of transformations.

What? While the statement may seem a bit technical, it represents a set of clear notions that explain, among other things, what geometers really mean by such expressions as "only one" or "the same." Klein's definition of geometry is a very rich concept and need a bit of getting used to. This chapter will be devoted to finding out what a part of Klein's definition of a geometry is all about. The full meaning will become clear after Chapter 8, when groups are discussed.

When Are Two Things the Same?

When you say *only one* triangle meets some given conditions, another triangle that matches the conditions is considered to be the same triangle as the first in some sense. In Chapter 1 we avoided the issue of calling two triangles that had sides with each of the lengths the same by talking about congruence. Congruence was used as an undefined concept and interpreted as the ability to move one figure so that it could be superimposed on another. Superimposition—or moving two figures so that they coincide—is a good idea, but it is not really *mathematical.* A mathematical idea is one that can be incorporated into one of the games that mathematicians play. Recall that many mathematicians think that their subject is just a set of arbitrary rules, as chess is. But these rules are very specific, and picking up figures from a plane and moving them about in space must be pinned down so that it can be described in tedious detail. We will avoid doing that, but just sketch out the main steps along the way.

In Chapter 1, the notion that symmetrical figures are congruent was stated as one of the rules of the game. We can keep that rule and add another. *For any set of points and any line in a plane, you can create another set of points that is symmetrical to the first about the line.*

This rule makes sense if you picture a plane that can be folded at any line lying in the plane and a figure as something made from wet ink that will transfer to the other side; or if you treat the mirror image of a figure as real. But both of these ways of thinking about the new rule suggests that you end up with two figures, not one figure shown twice. Furthermore, neither way is really suitable to mathematical treatment.

There is another image of the rule that is more useful in sorting out the meaning of "the same." Picture the plane as transparent, so that you can see the figure from either side. Instead of folding the plane about a line of symmetry, the line of symmetry can be taken as an axis about which the plane can be rotated. When the plane is rotated about the axis, the symmetrical figure will come into view, although from the other side of the plane, just as stated in the rule. Mathematicians who use this image prefer to call line symmetry *axial symmetry* because they view the line as an axis like the axis from the north to the south poles about which the Earth turns. Another way to state the new rule is that it is always possible to rotate the plane about the line that you choose and to see the figure from the other side. Now you can define two figures to be the same when one of them is formed by this operation from the other. This definition is even more convincing than talking about folding planes or mirror images because the figure really *is* the same figure, just looked at another way. Perhaps an even more finicky way of thinking of this would be to keep the plane stationary and walk around it yourself, but attractive as this thought is, a walk around the plane is hard to describe in mathematical terms.

Since the figure after performing the operation is the same figure as before rotation, just viewed from the other side of the plane, it makes sense to say that this is the basic definition of when two figures are the same. The operation itself is called *reflection.* Therefore, you can say that two figures are the same when one is a reflection of the other. Of course, there is a sense in which you mean two different figures and another sense in which you really means the *same* figure. Perhaps a better way to say it would be "same for all practical purposes." In fact, if the original figure was asymmetric, then the reflection does not even look like the original figure. Remember that MOW becomes WOM.

Clearly, however, you want two figures that do look exactly alike also to be called the same. The solution to this is relatively simple. Reflect the original figure twice, either about the same line or about two different lines. This is in some ways like looking at the figure from one side of the plane, walking around the plane and looking at it from the other side, then coming back to the first side to look at it again. If the reflection is about two different lines, the corresponding result for walking round the plane might be standing in a different spot when you get back to the original side of the plane. In any case, it is clear that not only is the figure "the same" in these operations, but it even looks the same when you get back to the original side.

At this point, it is probably a good idea to begin to use a technical term for "the same," since this phrase has been given a very specific definition that is not always consistent with its informal use. Mathematicians call properties of geometric figures (and other ideas, as well) *invariant* when they are the same in the sense discussed here. That is, a property is invariant if the property is not changed by an operation that is performed on it. The property of being a line segment or of being a circle is invariant with respect to the operation of reflection about a line.

The property of distance between points or the property of the size of an angle is also invariant with respect to the same operation. Also, if two lines are parallel in the original figure, they will be parallel in the reflection, so parallelism is invariant for reflection. And so forth.

What was it Felix Klein called geometry? Geometry is the set of properties that are *invariant* under a particular group of transformations. We are getting a bit closer.

But, you might ask, since the reflection in this definition is really the same figure as the original, are not all properties invariant under reflection? Reflection would hardly be worth thinking about if they were. Remember that asymmetrical figures do not look the same, so clearly something is changing. Also, consider what happens when you reflect an arrow across a line (Figure 3–1).

The original arrow is pointing toward the left, but the reflection is pointing toward the right. Something is not invariant. You can call the property that is not invariant *orientation.* The orientation of an arrow is the property of pointing in a given direction. It changes for reflection unless the "shaft" of the arrow is parallel to the axis.

You can get back to the original orientation by re-reflecting the arrow about the same axis, which returns you to the original position of the arrow in the plane. The combined operation of reflecting twice about the same axis is important theoretically, as you will see later, but practically it does not get you anywhere. (This is the operation for which *all* properties are invariant that was just described as "hardly worth thinking about.") What is

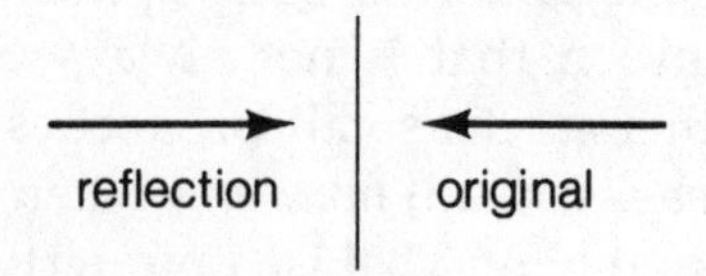

FIGURE 3–1
The property that is different about the original arrow and its reflection is called ***orientation.***

needed is an operation that leaves the arrow in a different place on the plane, but keeps orientation invariant.

Mathematicians like to keep the number of assumptions as close to the minimum as possible. Rather than introducing a new operation, it is preferable to use the one we already have. Since reflection twice about the same line is unsatisfactory, look again at what happens after two reflections about different axes (Figure 3–2). A little experimentation should show you that there are two different situations.

In A, the orientation is lost in both reflections, but in B the arrow comes back to its original orientation after the second reflection. What is the difference between the two situations? Informally, it is easy just to say that the axes in B are parallel and the axes in A are not.

This still leaves *orientation* undefined. Basically, orientation is what is invariant in two reflections about parallel lines. Reflection works like multiplying by a negative number. One multiplication changes the sign and two multiplications keep the sign the same. One reflection changes the orientation and two reflections (in parallel lines) keep the orientation the same.

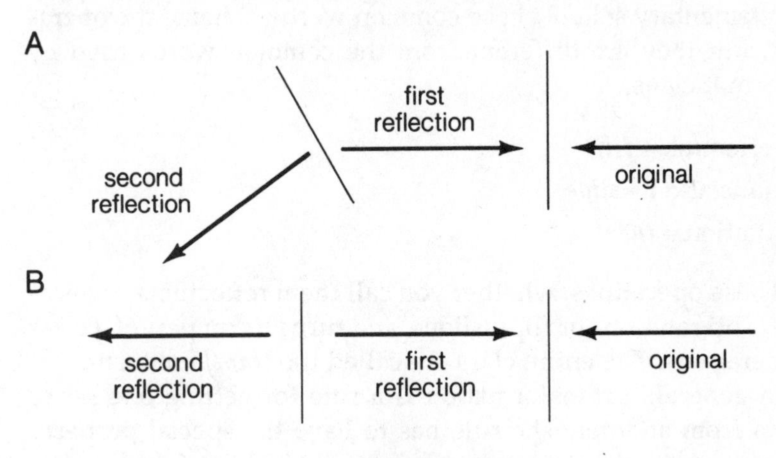

FIGURE 3–2
When two lines are parallel, reflecting once about each restores the orientation, but when they are not parallel, a different orientation results from double reflection in the lines.

Since this operation of reflecting in two parallel lines has this useful invariance, it is a good idea to give the operation a name. The name is *translation,* which may seem odd if you think of translation from one language to another. When you translate a geometric figure, nothing about it changes except for its position in space, but when you translate a word from one language to another, generally, everything about it changes but its meaning. The root of the word *translate* means "carried across," so it is easy to see how the two different senses of the word evolved.

The double-reflection about two lines that intersect has a name as well. This operation is called *rotation.* The reason for the name becomes clear if you note that the operation of double-reflection about intersecting lines has the same result as turning the plane about the point of intersection by twice the angle formed by the lines. Similarly, you could also think of the operation of translation as simply sliding the plane along a path perpendicular to the two parallel lines for twice the distance between them (Figure 3–3).

Sometimes these same three operations are taught in elementary school. The textbook authors who introduced the operations into elementary school chose common words to name the operations, but they are different from the common words used by mathematicians.

reflection = *flip*

translation = *slide*

rotation = *turn*

These operations, whether you call them reflections, translations, and rotations or flips, slides, and turns, form part of a class of operations. The entire class are called the *transformations.*

In general, a transformation is a rule for getting one set of points from another. The rule has to have the special property that if you *transform* any one of the points in the first set, you will get one and only one point in the second set. Furthermore, it is possible to work backwards. Given a point in the second set, there is one and only one point to which it corresponds in the

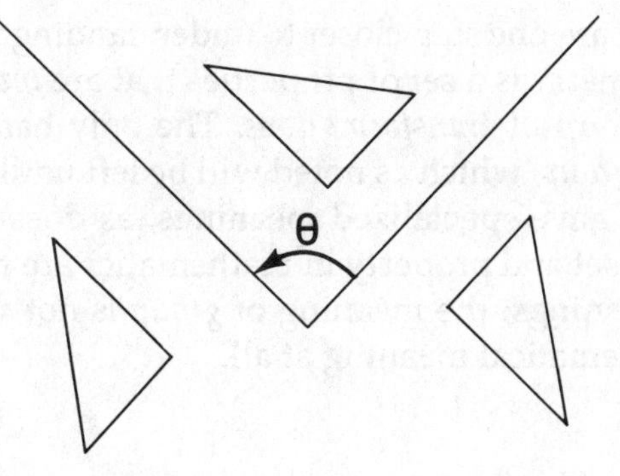

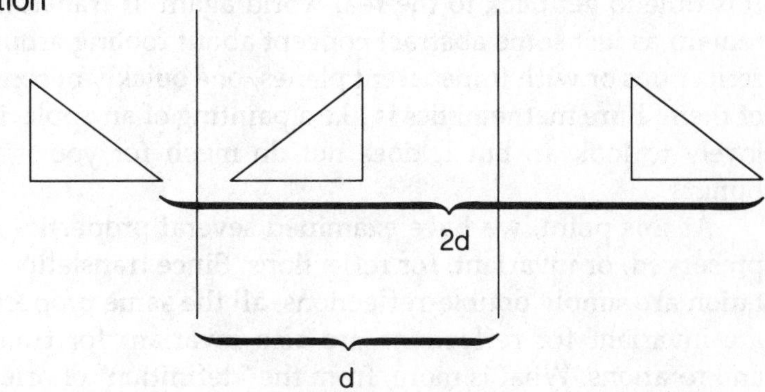

FIGURE 3–3
Reflection in two nonparallel lines results in a rotation through twice the angle between the lines, while reflection in two parallel lines results in a translation of twice the distance between the lines.

first set. This set of requirements is called *one-to-one correspondence.* Therefore, a *transformation is a one-to-one correspondence between two sets of points.*

It is not necessary that all the points of the first set be different from all points of the second. In fact, all the points can be the same, as in the example of double-reflecting a figure about just one line. This transformation also has a name—the *identity* transformation.

Now you are one step closer to understanding what a geometry is. A geometry is a set of properties that are *invariant* under a particular group of *transformations.* The only hard word left to deal with is *group,* which as noted will be left until Chapter 8. *Set* and *property* have specialized meanings, as does *group.* But the meanings of set and property in mathematics are not unlike their ordinary meanings; the meaning of *group* is not very much like the nonmathematical meaning at all.

What's the Use?

It is time to get back to the real world again. If transformations remain as just some abstract concept about fooling around with reflections or with transparent planes, one quickly becomes tired of them. Pure mathematics is like a painting of an apple; it can be lovely to look at, but it does not do much for you if you are hungry.

At this point, we have examined several properties that are preserved, or invariant, for reflections. Since translation and rotation are simply double-reflections, all the same properties that are invariant for reflections are also invariant for translations and rotations. What is more, from the "definition" of orientation, it is necessary that orientation be invariant for translation (orientation is that property that is invariant for translation but not for reflection).

The Bridge of Tran Slate Way

Two towns are situated on either side of a giant, straight canyon. For years, the people from Tran had nothing to do with the people of Slate because of this impediment between their towns. Finally, the government decided to build a road between the towns. Since the land between the towns is flat except for the canyon (in fact, the people in Tran can see the town of Slate, although it is a few kilometers away), the road could be perfectly straight—except for the canyon. The canyon needed to be bridged, and the economics of the situation were that the bridge had to be perpendicular to the canyon walls, which would make the shortest, and least expensive, bridge. But the roads from each town to the bridge also should be kept as short as possible. Where was the best place to put the bridge to minimize the length of highway to be built?

Transformations to the rescue: Begin by assuming that you have solved the problem. If the towns are called *T* and *S*, then the bridge is built where the line segment *AB* is in Figure 3–4.

Translate the bridge so that point *B* coincides with point *T*. The translation of *AB*, which you can call *A′T*, has some properties the same as *AB*—namely, those properties that are invariant for translation. In this case, they are the length of the line segment ($AB \cong A'T$) and the orientation of the line segment ($AB \parallel A'T$). Furthermore, if you connect *A* and *A′*, the line segment formed will also be a translation of *TB*, and therefore congruent and parallel to *TB*. (In general, if the two endpoints of a segment are translated the same distance in the same direction, the segment between the translated points will be a translation of the original segment.) Now you have the situation shown in Figure 3–5.

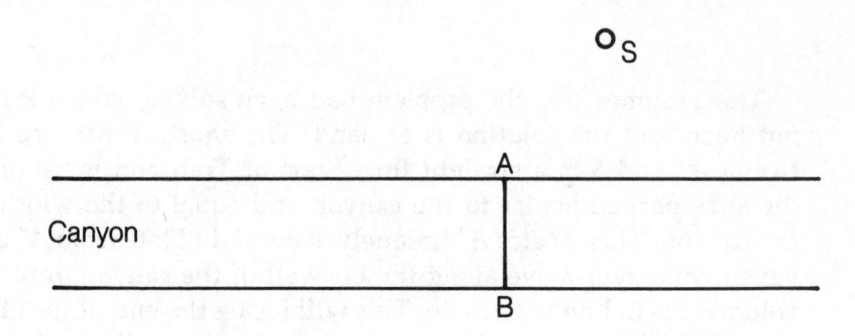

FIGURE 3–4
Assume that the line segment AB is the bridge across the canyon between Tran and Slate.

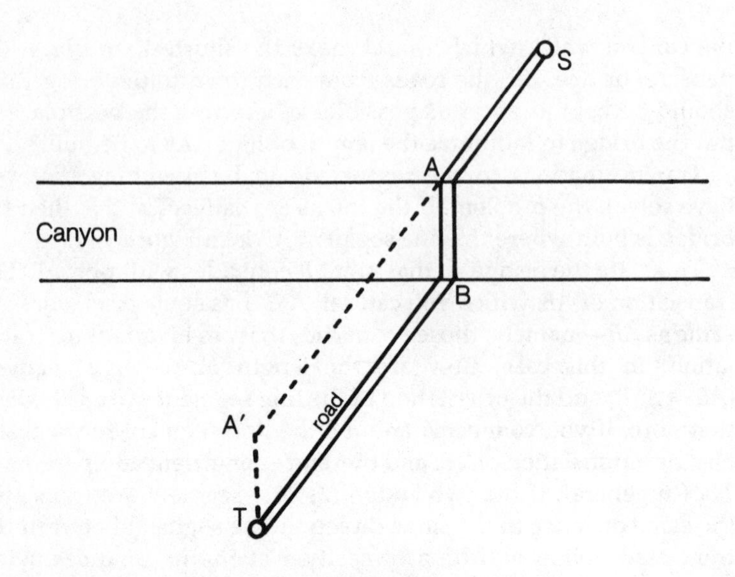

FIGURE 3–5
To solve the problem, AB is translated to the position A′T and the line from A′ to S determines where A should be placed.

This assumes that the problem had been solved, which it had not been. But the solution is at hand. The shortest distance between *A′* and *S* is a straight line. Start at Tran and mark off a distance perpendicular to the canyon and equal to the width of the canyon. This locates *A′* uniquely. Now sight Slate from *A′* and have a volunteer move along the far wall of the canyon until the volunteer is in line with Slate. This will locate the end of the Slate side of the bridge. The Tran side of the bridge will be directly across from it. Roads can go straight from Tran and Slate to the bridge, giving the minimum possible cost for the Tran-Slate Way.

There are not a whole lot of similar "practical" problems that can be solved by the limited number of transformations that we have at hand, but the general technique is extremely useful. The idea is to use an appropriate transformation to turn a problem that you cannot solve (or would have difficulty solving) into a problem that is easily solved or for which you already know the solution. A common example that is encountered in school is handling the graphs of the conic sections (the circle, the parabola, the ellipse, and the hyperbola). First you learn how to graph the simplest conic sections—the circle with its center at the origin of the graph, the parabola that is the graph of $y = x^2$, and so forth. Then, if you are faced with a circle whose center is not at the origin or with some other parabola, for instance, you use the appropriate transformation to put it right where you want it.

There is an amusing essay that appeared in *The American Mathematical Monthly* about 50 years ago about how a mathematician or a physicist would trap a lion in the middle of the Sahara. It was written by a mythical mathematician named E. S. Pondiczery who, although nonexistent (he was the pseudonym for a group of mathematicians), used another pseudonym, H. P'tard, on the article. Different methods for trapping the lion are used by mathematicians or physicists who specialize in different branches of mathematics or physics, but the mathematician who specializes in transformations has one of the most effective. First the mathematician builds a spherical cage for the lion and gets inside the cage. Then he or she applies the appropriate transformation to the lion so that the lion is placed in the interior of the cage and the mathematician is placed outside. (Unfortunately, the transformation is of a type similar to reflection. Although not noted by the doubly pseudonymous author, the result would no longer be a lion or even what Alice would call a "looking-glass" lion. The consequences to the lion and to the mathematician are truly unpleasant, since the transformation, known as inversion, would certainly turn both of them inside out.)

Let a Tile Be Your Umbrella

If you consider a figure that undergoes the same translation or reflection a number of times, you have what is known as a *frieze pattern.* These are the kinds of patterns that are found today on wallpaper borders, but were used in the past in marble friezes around the tops of buildings (Figure 3–6).

FIGURE 3–6
Patterns such as these are commonly used as friezes in classical architecture.

With the introduction of frieze patterns, the idea of symmetry is generalized to include figures that are not easily recognized as symmetric from the experience one has in school. I know that this caused me great confusion when I first encountered the word *symmetry* used in this context, so I am going rather slowly in the hope that you won't have the same conceptual difficulty that I did. When I look at a frieze pattern, I do not automatically react "Ah, symmetry!" but that is the reaction you need to develop to be able to understand some of the important ways that symmetry is used in higher mathematics or in physics.

Take the easier case first. A frieze pattern based on repeated reflections is symmetric in the ordinary, elementary, left-right sense. Assume, however, that the pattern extends infinitely in both directions. If you could take the infinite pattern of reflections, there is a line about which you could fold the infinite strip so that the part on one side would exactly match the part on the other. In fact, since the pattern is infinite, there is more than one such line. All in all, there are an infinity of lines of symmetry.

Now look at the more difficult case, a frieze pattern made from translations. The key lies in the question we asked in Chapter 1 about symmetry: "What can you say about the subject after you've said one part looks just like the other?" Mathematicians and scientists have agreed that when one part looks just like the other, they will use the word *symmetry* to describe the situation even when one part is not a mirror reflection of the other. In the case of a frieze made from translations, the part is repeated over and over again in a long line. Although this seems to have little to do with line symmetry, it is a kind of symmetry in any case.

In this regard, *symmetry* is unlike any other word in mathematics because it is defined at more than one level. Level 1 is the definition of line symmetry presented in Chapter 1 and also point symmetry, which is discussed later in this chapter. Level 2 is the meaning that is being developed now; basically, that figures are symmetric if one can be transformed into another by a transformation for which congruence is invariant. Level 3, which is related to numerical symmetry, is developed in Chapter 5. Level 4, which may be termed an algebraic definition, is the subject of Chapter 8. Level 5, which may be called "perfect symmetry," is discussed in Chapter 9. All of these different levels, however, are not different meanings of the word; rather they are *extensions* of the basic idea, each of which subsumes the lower levels in some way. *Symmetry* is not like *base,* another word that seems to have different meanings in mathematics. The base angles of a triangle, the base of a numeration system, and the base of a power have little or nothing to do with each other. All the meanings of symmetry have everything to do with each other. It is possible to talk about degrees of symmetry in such a way that as a substance develops (such as a crystal forming) what you would at this point recognize as symmetry, from a more advanced point of view the substance is actually losing symmetry.

Let's go back to the symmetry of a frieze pattern formed by translation. Since the basic figure (choose any one of them) is repeated by translation, each of the figures is congruent to all the others. "One part looks just like the other." In that case, you can say that the strip as a whole has symmetry—translational symmetry.

Frieze patterns, while strictly speaking two-dimensional, are essentially one-dimensional in the sense that they do not extend from their strip to cover the plane. In fact, the geometer H. S. M. Coxeter has called them "1 1/2 dimensional." But it is easy to extend the idea of either reflective symmetry or translational symmetry to the entire plane or to three dimensions (or more; mathematicians never like to stop at three). That is, it is easy to get the general idea in two or three dimensions; the complexity of such patterns in more dimensions is quite beyond the scope of this book.

There are only seven essentially different ways to generate frieze patterns (of which you have encountered two so far). One pattern, of course, is translational symmetry, which can be illustrated with an infinite row of F's. F is chosen because it is a letter with no symmetry of its own.

. . . F F F F F F F . . .

A second pattern can be thought of as the pattern generated by reflecting an object such as W which has line symmetry in two parallel mirrors. It can be also be represented with an infinite row of a letter of the alphabet, such as

. . . W W W W W W W . . .

Yet another pattern involves symmetry about a horizontal line combined with translation. This has more symmetry than just translation, since the basic figure that is translated has more symmetry.

. . . C C C C C C C . . .

If the figure is symmetrical about both a horizontal and a vertical line, you get the fourth basic pattern.

. . . H H H H H H H . . .

The fifth one can be best thought of in terms of a symmetry that we have not dealt with as yet, but will get to in the next section. A peek at a representation won't hurt, however. We lack the necessary combination of letters in the Roman alphabet, however, so I'll have to borrow one from Greek (the capital gamma).

. . . L Γ L Γ L Γ L . . .

Other patterns use yet another form of symmetry, alone or in combination with other symmetries.

. . . N N N N N N N . . .

. . . W M W M W M W . . .

You will also encounter this form of symmetry later in the chapter.

The situation gets more complicated when one goes to two dimensions, as we are about to do. There are seventeen different ways to generate the two-dimensional version of a symmetrical frieze pattern, and they include symmetry operations that we won't consider until nearly the end of this book.

The basic notion of going to two dimensions is just to translate some figure repeatedly, but not to limit the translation to a single direction (Figure 3–7).

Again, you can say that the figure (taken as a whole) has translational symmetry. If the triangle had been an isosceles

FIGURE 3–7
When a triangle is translated by equal amounts in two perpendicular directions, it illustrates a generalized form of symmetry.

triangle, a figure that by itself has line symmetry, then the entire pattern would have more symmetry than it does for the right triangle. If the figure had been the letter H, which has two lines of symmetry, there would still be additional symmetry to the pattern. This is one useful way of looking at symmetry in the plane.

Another useful way of looking at symmetry in the plane is to consider how the plane can be divided up into polygons that cover the whole plane without overlapping. One way to do this is to begin by looking at just a single point of the figure being translated and all the translations of that point. Say you choose the right-angle vertex of one of the right triangles. The set of all translations of that point is called a *lattice* (Figure 3–8).

The reason that this figure is called a lattice can be made clearer by connecting four of the adjacent lattice points, which will form a parallelogram. Now translate that parallelogram in two directions at once so that its image is adjacent to the original parallelogram. Repeat the operation to cover the entire plane with such parallelograms. The result looks like the kind of lattice that is sometimes used as a trellis for dividing rooms or growing

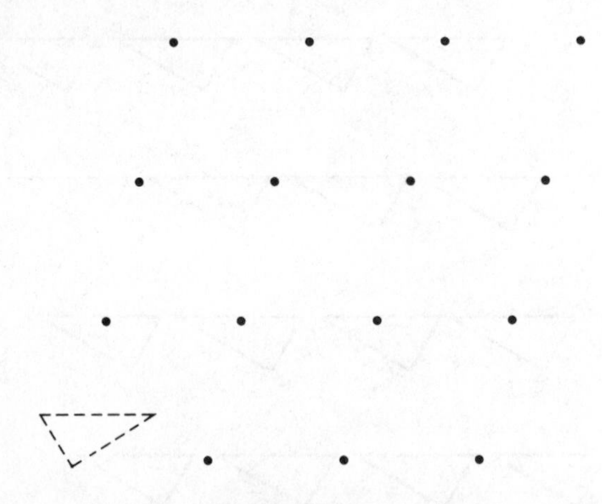

FIGURE 3–8
One point of the triangle that is translated by equal amounts in two perpendicular directions forms a mathematical structure called a *lattice*.

vines (Figure 3–9). This method provides one way to divide the whole plane into polygons that cover it without overlapping.

Any pattern of polygons that covers the plane in this way is called a *tesselation.* The most common tesselations that you are likely to see around you are made from tiles, especially tiled floors. The tiled floor in my office is a tesselation formed from squares. Older bathrooms often have floors tiled with regular hexagons. (*Regular* for polygons means that all the sides and all the angles of the polygon are congruent; a square is a regular polygon, but a rectangle that is not a square is not regular and a rhombus that is not also a square is not regular.)

Other cultures also made use of tiled patterns of varying degrees of complexity. For example, the Moors of Spain were strict Moslems, who took seriously the implications of God's Second Commandment: "Do not make for yourselves images of anything in heaven or on Earth or in the water under the Earth." As a consequence, they developed a fantastic nonrepresentational art style based on geometric forms, which are particularly suited to tiling. In fact, in the Alhambra, all 17 of the two-dimensional translational symmetry groups are used.

The ancient Greeks also used tiled patterns. Some have speculated that Pythagoras (or one of his predecessors) discovered the well-known Pythagorean theorem by contemplating a tiled pattern of isosceles right triangles. It is as easy to tile a floor with isosceles right triangles as it is to tile a floor with squares. In fact, if you cut a diagonal along each square of the tiles on my office

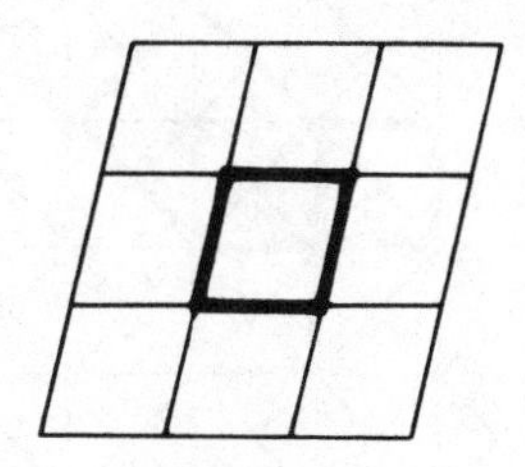

FIGURE 3–9
A parallelogram can be translated in such a way that it generates a set of congruent parallelograms that cover the plane.

floor, the floor will be completely covered with nonoverlapping right triangles. By rejoining the triangles, you can obtain a tesselation into squares but there are two different ways to do this. One way produces the original tiled floor, but you can also join four triangles at their right-angle vertexes to form a larger square. If you look for a long time at a floor tiled in isosceles right triangles, you will find that your attention wanders back and forth between seeing the pattern of small squares and the pattern of large squares (Figure 3–10).

It is somewhat amusing to picture an ancient Greek bathroom floor tiled with isosceles right triangles and old Pythagoras in the room for some extended period of time on some mission or other with nothing to occupy his fertile mind. He stares at the floor; first he notices the small squares, then he notices the large squares, then he notices the small squares; and so on. Suddenly he puts the two images together in his mind: The square on the hypotenuse of the right triangle is equal to the sum of the squares on the other two sides! Violà, the Pythagorean theorem!

The square on the longest side (the *hypotenuse*) is formed from four right triangles, while the squares on each of the two short sides are formed from two right triangles each. Since all the right triangles are congruent, the theorem is proved (for isosceles right triangles, only; but once you have the basic idea, you can go on to prove it for other right triangles).

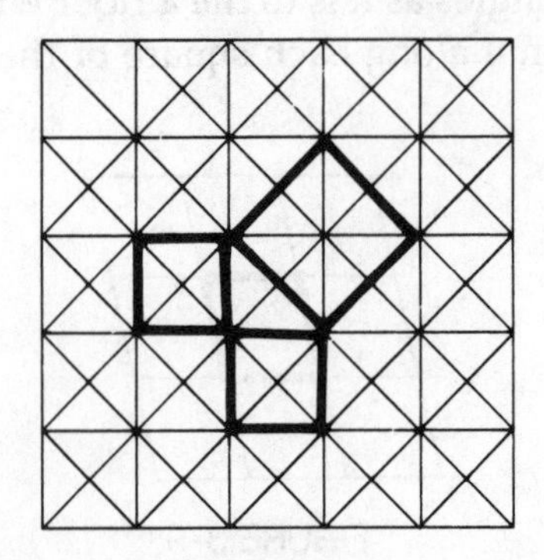

FIGURE 3–10
A tessellation of right triangles suggests the Pythagorean theorem.

This account of the discovery by Pythagoras is highly speculative, since it is likely that the famous theorem was actually obtained from the Babylonians, who had it hundreds of years before Pythagoras. The earliest known Chinese mathematical documents also contain figures that suggest that the Chinese knew at least the Pythagorean theorem for isosceles triangles, and they illustrate it with a drawing that looks much like a tiling of isosceles right triangles.

Eschervescence (Part 1)

Tesselation of the plane of a sort can be carried out with figures that are not simple geometric shapes, as some of the figures in the Alhambra demonstrate so well. But in a culture that does not take the Second Commandment seriously, the scope of such tilings can be extended to include representations of any forms, including even those creatures that probably exist only in our imaginations. One serious artist of our time has taken advantage of this to produce works of art that are a particular delight to mathematicians and scientists—M. C. Escher (1902–1972).

When Escher began filling planes with strange and familiar creatures, he knew nothing of the mathematics behind his work. In fact, his first "periodic drawing," as he called this aspect of his work, was made in 1922 and it was not until 1935 that he learned that mathematicians had worked out a complex theory for such drawings. When he learned that mathematicians and crystallographers had done remarkably similar work to what he had been doing, he studied some of their ideas, but he continued to find his own concepts sufficiently fertile without drawing upon those of others (except perhaps for his paradoxes, which derive from the work of mathematician Roger Penrose. We will not deal with them in this book; if you are interested in the paradoxical paintings and lithographs, I recommend *Gödel Escher Bach: An Eternal Golden Braid* by Douglas R. Hofstadter published by Basic Books, New York, 1979).

Figure 3–11 is an Escher periodic drawing of the simplest possible type, showing the translation of two figures to fill the plane.

Other Escher drawings involve symmetry operations (that is, transformations) that have not been discussed so far. For example, Figure 3–12 appears at first glance to involve simple reflection, but further consideration shows that the situation is more complex.

If, say, the white man is considered, the transformations of the man looking to the left show us his left side with left arm outstretched, while those shown looking to the right portray the right side with the right arm outstretched. The transformation is some sort of reflection, even though there is no place where you could insert a mirror to get the image. Mathematicians call this transformation a *glide-reflection,* an awkward but descriptive name. A common way to explain a glide-reflection is to say that it is a translation followed by a reflection. In this book, however, it makes more sense to say that a glide-reflection is the result of

FIGURE 3–11

In M. C. Escher's *Frog and Fish*, the plane is covered by translations of two different creatures. © 1989 M. S. Escher Heirs/Cordon Art—Baarn, Holland

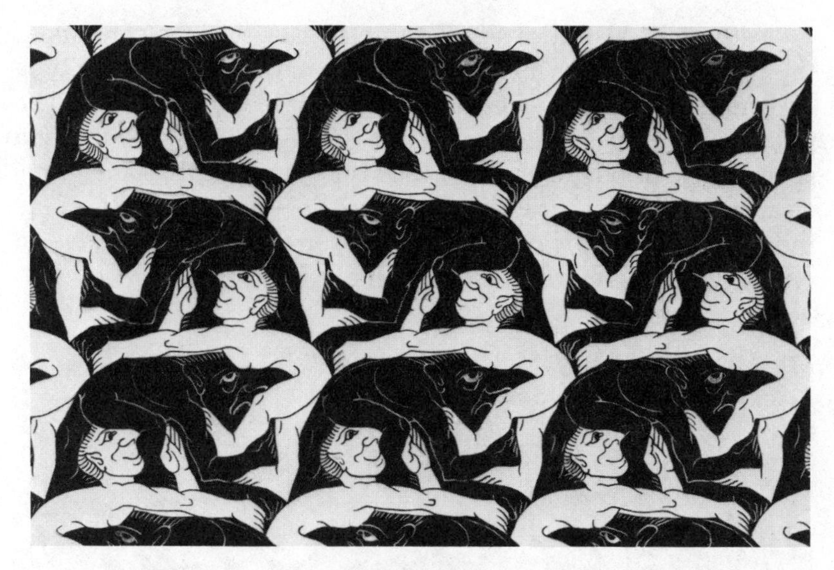

FIGURE 3–12
Escher's *White Man, Black Man* illustrates a different transformation, glide-reflection, which is a cross between translation and reflection. © 1989 M. S. Escher Heirs/Cordon Art—Baarn, Holland

three reflections about three different lines, of which two are parallel and the third is not. In that way, all the obvious invariances of reflection are established.

With the introduction of glide-reflections, you now have all of the transformations of the plane that leave congruence invariant: reflection, translation, rotation, and glide-reflection. Each of these can be considered a symmetry operation as well. Other transformations—those that are not invariant for congruence—are not symmetry operations. An example of a transformation that is not a symmetry operation is *dilation,* the expansion or contraction of the whole plane. The result of dilation is a figure that is similar to, but not congruent to, the original figure. Since symmetry means that the parts are exactly alike, this sort of transformation destroys symmetry in the sense that it has been discussed so far. At one of the higher levels of understanding, however, dilation can be interpreted to preserve some forms of symmetry. For now, however, let us stick to the level before us.

Although the entire set of symmetry operations consists of the four congruence-invariant transformations, one other type of symmetry and one other type of transformation can be pointed out. This transformation is, on the one hand, simply a special case of one of the first four types and, at the same time, clearly a common type of transformation that can be defined independently of the other four. It is conveniently illustrated in one of Escher's periodic drawings (Figure 3–13).

FIGURE 3–13
Escher's *Birds and Fishes* introduces an important special case of rotation, the transformation known as reflection in a point. © 1989 M. S. Escher Heirs/Cordon Art—Baarn, Holland

Focus on the white birds, which stand out more (and are a bit more realistic) than the dark fishes. Some of the birds could have been formed by translation from one of the others, but this accounts for only half of the birds. The other half require some other kind of transformation. Again, the transformation has a lot in common with reflection, but it is not pure reflection in a line. To superimpose one of the white birds on one of the birds that cannot be reached by translation, the simplest way is to rotate the plane about a point exactly 180 degrees. This transformation can be termed a *half-turn* or *reflection in a point.* If it is defined in terms of ordinary reflection (that is, reflection in a line), then reflection in a point consists of reflection twice in intersecting lines (that is the general definition of rotation) that are perpendicular to each other (the condition that makes the rotation exactly a half turn; since the angle rotated is twice the angle between the intersecting lines, the angle rotated as a result of two reflections about perpendicular lines is $2 \times 90° = 180°$). Reflection in a point is worth spending some time on.

Reflection in a Very Small Mirror

In Chapter 1, you were given a problem to work out on your own. As you examined the letters of the alphabet for line symmetry, in addition to letters with vertical symmetry, such as A, and letters with horizontal symmetry, such as B, it was pointed out that such letters as N seem to have a sort of symmetry that is not line symmetry. Other letters that share this property include S and Z. You were challenged to use the simple concepts in Chapter 1 to develop a definition of this form of symmetry. Here is the solution to that problem.

The parallel idea to reflection in a line is reflection in a point. When defined independently of reflection in a line, it is actually simpler than reflection in a line. For this reason, reflection in a point is taken as basic in some systems, and reflection in a line derived from it. The only concepts required are *point* and *distance. One point is the reflection of another in a specified point if the original point and its image are the same distance from the specified point and all three points are on the same line.*

As before, the definition can be extended to all kinds of plane figures. When all the points in a figure are reflected in the same point, the resulting set of points is the reflection of the original figure in the given point. If you reexamine the alphabet once more, you can observe that some letters that have line symmetry also have point symmetry, specifically H, I, O, and X. The key operation for checking reflection in a point is simply turning the page upside down, since, as previously noted, reflection in a point is the same operation as a half-turn.

Some possible confusion about symmetry may come from the fact that symmetry is both an operation (that is, a transformation) and a condition. For example, you can look at a figure, such as one of the letters of the alphabet and say "this figure has line symmetry; or this figure has point symmetry." In that case, symmetry is being used as a condition or description of the figure. On the other hand, while a figure such as the letter F has no intrinsic symmetry, it can be reflected about a line or in a point (as can any other figure), which is one of the symmetry operations.

These ideas can be related by counting the number of symmetry operations that leave a figure unchanged (all properties invariant). The number and type of operations that leave the figure invariant characterizes the kind of symmetry that the figure has. In this case, we are talking primarily about a description of the figure in terms of an operation. For example, a letter such as A or B, which each has one line of symmetry and no point of symmetry, can be carried into itself in only one way—by reflection about the single line of symmetry. However, it is also customary to count the identity transformation as a symmetry operation, so the symmetry of A or B is described as being *order* 2 symmetry. Similarly, N, which can be converted into itself only by a half-turn or by the identity transformation, also has order 2 symmetry, but H, with its two lines of symmetry and point symmetry as well, has order 4 symmetry.

It is natural to wonder about orders other than 2 and 4. If a figure cannot be transformed into itself in any way except by the identity transformation, the order is 1. Figure 3–14 is a figure with order 3.

This is an ancient magical symbol, known as the triquetrum. Although it has neither line symmetry nor point symmetry, it can be seen to have some kind of symmetry even by someone who has no clear notion of exactly what symmetry is. In this case, the symmetry operation is not a half-turn but a third-turn—a rotation of 120 degrees. Each third-turn will bring the figure into coincidence with itself, but three third-turns is exactly the same as the identity transformation. Therefore, a triquetrum has order 3. You might think that an equilateral triangle would also have order 3, but a little reflection (pun? —well, double-meaning intended) will show you that the equilateral triangle can be transformed into itself around 3 lines of symmetry as well as with 3 third-turns, so an equilateral triangle has order 6 symmetry.

Just as the order of a figure can be used to characterize its symmetry, the symmetry operations can also be characterized as different types. For transformations of the whole plane, the key characterization is in terms of the part of the plane that is left invariant by the operation. For example, a reflection about a line changes every point in the plane except for the points on the line. Here you must picture the plane rotating around an axis, which is the invariant line. A rotation of any kind (including a half-turn) amounts to two reflections about intersecting lines, so the only point on the plane that is not changed is the point of intersection

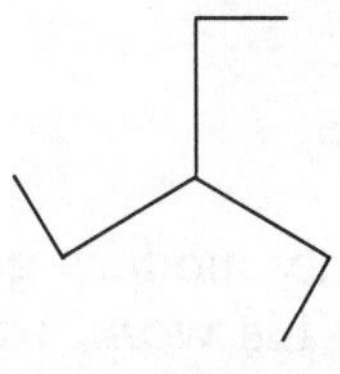

FIGURE 3–14
A triquetrum is a figure that can be transformed into itself by rotation through 120°, or a third of a circle.

of the two lines. Another way to look at this is that if the plane is thought of as being balanced on the point of a needle, a rotation consists of turning the plane about that balance point. Every point changes except the balance point.

For a translation or a glide-reflection, a reflection about two parallel lines is involved. Parallel lines have no points in common, so no points of the plane are invariant in either of these two transformations.

The transformations can also be characterized by whether they leave the orientation of a figure the same or change it. As noted earlier, a single reflection in a line *changes* the orientation, two reflections in parallel lines change it back, and a third changes it again. In other words, reflection changes the orientation, while translation does not. There is even a more fundamental property that changes in a way similar to orientation. It is called *sense.* If you have labeled the vertices of a triangle, you can think of the sense as the order of the labels. When a triangle is reflected about a line, the sense of the triangle changes even if one of the sides is parallel to the line of reflection. If the triangle is not symmetric, then it cannot be slid along the plane to coincide with its reflection. Glide-reflection also changes the sense of a figure, while neither translation nor rotation changes the sense. Transformations that change the sense are called *opposite* transformations, while those that do not change sense are called *direct.*

Both the idea of orders and of sense can further illuminate the classification of types of friezes.

Music to My Enantiomorphic Ears

Although Escher is the only modern graphic artist to use transformations extensively in his work, transformations of another kind have long been used in another art form. In Escher's work, the transformations are of the plane, so they are exactly the same as the geometrical transformations. The concept of transformation can be extended to transformations in other dimensions, such as music.

The dimensions of music are time and pitch instead of being left-right and up-down, as on a plane. Music has other characteristics such as harmony and timbre, but these can be ignored for now, just as we are ignoring for now any characteristics of geometric figures other than their shape and size. Most of the transformations of the plane have exact analogs in music and have been used extensively by composers, especially by the composers of the seventeenth and eighteenth centuries. Bach is to music what Escher is to graphic arts, a statement made quite clearly in Douglas Hofstadter's *Gödel Escher Bach,* although Hofstadter was primarily concerned with other aspects of the similarity of the work of Bach and Escher than we shall be here.

The easiest transformation to recognize in music is a translation in a single direction in time or pitch, the equivalent of a frieze pattern since such a transformation is in one dimension only. A translation in pitch is simply a change in key. Just as you recognize a triangle as being essentially the same no matter where it is placed on the plane, you also recognize a tune as being the same no matter what key it is in. If you know a tune, you will recognize it if it is played again, so tunes are invariant for translation in both time and pitch.

Bach exploited the invariance of a tune in pitch by writing music that ended up in a different key from which it started. When played through several times in a row (as happens in his composition), it eventually returns to the same key.

A familiar example of a translation in time is a round, such as *Row, Row Your Boat* or *Hi Ho Nobody at Home.* The tune is sung by several persons, each one entering the round at a later time. If the tune was written with this in mind, the result can be a pleasing musical experience.

A round is the simplest example of a musical form called the *canon,* which is based largely on transformations. Composers of canons also use the other transformations of time and pitch as well. For example, the two translations of time and pitch can happen simultaneously, so that the same tune re-enters the canon a few bars later as another voice but in a different key (while the tune also continues in the original key).

A more complex transformation rarely used in music is a half-turn (or reflection in a point). The comedian-musician Victor Borge occasionally uses this idea by announcing that he will play *The Blue Danube* and then playing something that sounds very strange, yet distantly like "Oh Danube so blue, so blue, so blue." He apologizes that he has accidentally put the sheet music upside down, turns the music upright, and plays a few bars of *The Blue Danube* in the normal way. Mozart, it is said, wrote a melody that is invariant for reflection in a point. The pitch intervals were exactly the same amount starting from either end of the piece, but in the opposite direction. Therefore, the same melody resulted when the sheet music was turned upside down as when the music was rightside up. Mozart's feat apparently is apocryphal, but modern composers have duplicated it, so it can be done. The result was not on the classical Hit Parade.

Use of a half-turn is unlikely ever to become a common device for the canon form. On the other hand, it is quite common for composers of canons to use the line reflection of the melody —that is, the same melody played backwards in time. Such a transformation in music is called a *crab canon* (although I think of a crab as moving sidewise, not backwards, so I would expect a crab canon to be a quarter-turn transformation).

A glide-reflection occurs less often. A glide-reflection in music would be a tune played backwards in time (or with the pitch changed so that everywhere the old pitch went up, the new one goes down) that enters the canon later than the original tune, and in a different key. Although not common, this kind of complexity has been tried also.

The analog to a rotation in music is hard to find in any pure form, but composers of canons frequently use a transformation that is not a symmetry transformation—dilation. That is, a canon melody may be repeated at a different tempo, which corresponds to a dilation in time. Dilation can be (and is) combined with the other transformations to produce the complex effects of the canon form.

If You've Got the Time and the Place

While symmetry in time is not a purely geometrical concept, it is of great importance in applying mathematics to the real world. If most physical laws or operations were not symmetrical, the world would be a far different place. As noted in Chapter 1, the reason that animals have left-right symmetry (and plants have what we can now recognize as a rough rotational symmetry) is that there is no reason for them not to have such symmetry. In the absence of forces to prevent symmetry from developing, it happens. Basically, however, the kind of symmetry involved is cognate to symmetry for figures, not symmetry for operations.

Even more important to science, however, is the kind of symmetry that we have been discussing in this chapter, symmetry for operations. For example, symmetry for translation and rotation on the surface of the Earth is accepted without thinking much about it. That is, if you carry an object from one place to another, you expect that it will still have the same shape and size in the new place as it did in the old. (Reflection and glide-reflection for real objects do not happen outside of Lewis Carroll or a few science-fiction situations.) Furthermore, it is clear that physical laws do not change for translation. Although gravitational force may vary over the face of the Earth, the law of gravity remains the same. In fact, all physical laws are invariant for translation on the surface of the Earth.

Newton and his predecessor Galileo made the first steps in understanding the importance of symmetry operations for physical laws. When Newton grasped the idea that an apple falling and the Moon falling must obey the same laws, he was merely extending the idea that translation does not change physical laws. Instead of just considering the surface of the Earth, Newton included all of space.

The recognition that physical laws remain the same throughout space became a major example of the symmetry of the universe as well as an important principle in developing new physical laws. A physical law that broke symmetry was highly suspect.

Symmetry for time is equally important. If a physical law worked one way on Mondays and a different way on Thursdays, it would not make for a very easy world in which to live. The principle of symmetry in time can be used in various ways. One of the first applications was in the field of geology, when James Hutton realized (in 1795) that the same processes that he could observe in the present as making small changes in the surface of the Earth could, over the course of many years, account for observed facts of geology. In other words, it was not necessary to assume that different processes were at work in the past to account for the present. This principle, known to geologists as "the doctrine of uniformity of process" or "uniformitarianism," was the major breakthrough in the development of geology, but it is essentially the observation that translation in time should not change physical laws.

Today, it is recognized that this symmetry is broken if you go far enough back in time; that is, there were different processes operating when the Earth was first formed than there are now. However, uniformitarianism continues to be a major guide to understanding the geology of the last two or three billion years.

The First Big Break

When I said that all physical laws are invariant for translation, I was not being quite accurate. It is true that physical laws are invariant for translation in a given space (which can be defined by the coordinates used as reference). For example, the laws of moving objects within a swiftly moving car are the same as those outside the car. If I gently toss a brick to you in the back seat, it will not cause much damage if you catch it. The same is true if we are both standing on the side of the road and I gently toss you the brick. But if you are standing along the road, and I gently toss it from a swiftly moving car, you had better not try to catch it! While the laws of motion are the same in each system, there seems to be some difference when you move across systems.

Similarly, the laws of physics we use daily all describe the space that is moving along with the Earth as it travels through that other thing we call space. That is, things that happen on Earth are seen with reference to a coordinate system that is moving along with the Earth. When you look at two different coordinate systems that are moving with respect to each other, the results are different. This difference is implied by a deeper understanding of symmetry that was developed around the beginning of the twentieth century.

Although the importance of symmetry considerations for physical laws was generally understood by scientists in the nineteenth century, no one focused full attention on them until Albert Einstein thought seriously about symmetry and invariance at the beginning of the twentieth century. Einstein specifically began with the question of what was invariant under uniform translation—that is constant translation of space, or uniform motion. Constant translation of space is the three-dimensional equivalent to considering one plane sliding over another at a constant velocity (velocity implying both speed and direction). He also considered what is invariant with respect to time in connection with a constant translation of space, for the first time linking space and time together in such a way that time can be considered as a fourth dimension. In his 1917 book for the general reader, he stated what he called "the principle of relativity (in the restricted sense)." This is just the symmetry idea we have been discussing, namely that if two space-times are moving with respect to each other in a uniform way (and not rotating), then "natural phenomena run their course . . . according to the same general laws" with respect to one space-time as they do with respect to the other space-time.

Having stated this quite clearly and having thought it over in depth, Einstein realized that it was not true. The symmetry is a broken one. As with other broken symmetries, what this implies is that events in the two different space-times obey the same physical laws—that is, show symmetry—as close as we can measure under "ordinary" conditions. Under "extraordinary" conditions, however, this sameness, or symmetry, is no longer true, and we are able to measure differences between the events in the two different space-times.

Einstein observed that the following two conditions, generally accepted as true by nineteenth-century scientists, must be rejected.

1. The time-interval (time) between two events is independent of the condition of motion of the body of reference.
2. The space-interval (distance) between two points of a rigid body is independent of the condition of motion of the body of reference.

Einstein's rejection of these ideas and his substitution of new principles for the absolute symmetry of space-time is the *Special Theory of Relativity,* first put forth in 1905. In the Special Theory, Einstein resolved the problem of invariance by showing that uniform motion of two coordinate systems (that is, uniform motion of one space-time with respect to another) does not result in translation after all. Instead, a different transformation, called a Fitzgerald-Lorentz transformation, is needed to describe such motion. The Fitzgerald-Lorentz transformation had been developed in 1889 by George Francis Fitzgerald and reintroduced in 1892 by Hendrik Lorentz. They used it to solve some difficulties that had arisen in physics as a result of the famous Michelson-Morley experiment which showed that light had a constant speed with respect to a moving frame of reference. Einstein showed that if one assumes that the speed of light is invariant in all steadily moving coordinate systems, both length and time obey the Fitzgerald-Lorentz transformation rather than the translation transformation.

Although the Fitzgerald-Lorentz transformations are not symmetry operations in the sense used in this chapter, they provide a kind of symmetry also. The symmetry is a result of the fact that conditions in each frame of reference are transformed in the same way by the Fitzgerald-Lorentz transformation. If, for example, you are in a space-time that is moving with respect to my space-time, I will observe that distances in your space-time are contracted by the Fitzgerald-Lorentz transformation in your direction of motion; but you will observe that distances in my space-time are contracted in my direction of motion. The same kind of equivalence holds for time. While this is a different kind of symmetry from the symmetry that results from translation, it is nevertheless recognizable as a symmetry concept. When Einstein broke the transformational symmetry of uniform motion, he replaced it with another type. The specifics of this change are presented in Chapter 5 after some necessary mathematical ideas have been developed.

After Einstein cleared the way, others began to think more seriously about the role of symmetry in physics. A particularly important result was obtained by the mathematician Emmy Noether (1882–1935) in 1918. She established that each symmetry principle in physics also implied a conservation law, such as the familiar law of conservation of energy—energy can neither be created nor destroyed.

Some of the principal quantities measured in physics include energy, momentum, and angular momentum, and there are conservation laws for each. You probably have a general idea of what energy is. Momentum is roughly a measure of inertia—the property of an object to keep moving if it is moving and to not move if it is not moving. Similarly, angular momentum is the measure of the property that keeps a top spinning (until frictional forces stop it). Noether's theorem relates such quantities to symmetry. For example, the conservation of energy is implied by translational symmetry in time. Similarly symmetry of space implies conservation of momentum, and symmetry of direction implies conservation of angular momentum. Other symmetries, as we shall see, imply other conservation laws for still other quantities.

One explanation of Noether's theorem is that symmetry always gives rise to invariance—by definition, the most general form of symmetry is that some properties are invariant. For the most commonly recognized symmetry, symmetry about a line, "one side looks just like the other." The figures on different sides of the line are invariant in some way. It is the invariance that produces the conservation laws.

Consider a particle in distant space, where there are no forces whatsoever acting on it. Suppose that the particle is at rest (with respect to whatever coordinate system you are using). Can it move to another point without a force acting on it? No, because if it did, it would have to choose a point to which to move—and all points are alike, because that part of space is perfectly symmetrical. If it did move, then its velocity would be variant, not invariant. Before it moved, it would be at rest. And conversely, while it was moving, it would not be at rest. Therefore, its momentum would change. But that cannot happen, because space is symmetric.

Similarly, angular momentum must be conserved because a particle in empty space could not choose one direction over another. If space were not perfectly symmetrical, momentum and angular momentum of an object would depend on the point in space at which the object was located.

Recently, some cosmologists have found evidence that could suggest that the universe as a whole is rotating. If this turns out to be the case, then it would be true that angular momentum and momentum would depend upon location in the universe. Just as experiments performed on a merry-go-round have different results if they are conducted near the outer edge than if they are performed near the center pole, a rotating universe would change everything. But for a small chunk of the universe—say a galaxy—the effect would not be noticeable.

Einstein's replacement of transformational symmetry with the symmetry of Special Relativity showed that the conservation laws of mass and energy do not hold under extreme conditions. A new symmetry law implies a new conservation law. The conservation law predicted by the Special Theory of Relativity is the conservation law of mass-energy. Energy alone is not conserved, for energy can be changed to mass according to the rule expressed in Einstein's well-known equation $E = mc^2$.

Summary

In this chapter, we have covered quite a lot of ground, starting with the question of why there is only one solution to the problem of constructing a triangle with three sides given and winding up with a new (in 1905, at least) theory of space-time. The connection of symmetry to art, music, and science is much more apparent than when the only symmetry notion was line symmetry—or Level One symmetry. These connections become even deeper and more meaningful as we advance to higher-level interpretations of symmetry.

4

Even an Odd Equation or Two

"I'm sure I'll take you with pleasure!" the Queen said. "Twopence a week, and jam every other day."

Alice couldn't help laughing as she said "I don't want you to hire me—and I don't care for jam."

"It's very good jam," said the Queen.

"Well, I don't want any today, at any rate."

"You couldn't have it if you did want it," the Queen said. "The rule is, jam tomorrow and jam yesterday—but never jam today."

"It must come sometimes to 'jam today,'" Alice objected.

"No, it can't," said the Queen. "It's jam every other day: today isn't any other day, you know."

Lewis Carroll

So far symmetry has seemed largely a part of geometry or else a part of physics, biology, or art related to geometry. True enough, but symmetry is also part of algebra. This may come as a bit of a surprise. It did to me when I first heard about it. Symmetry ideas exist in arithmetic, algebra, and function theory as well as in geometry. Although abstract symmetry is hardly as evocative as is symmetry in geometry, the concepts are ultimately more useful. Furthermore, when the various notions of geometric and algebraic symmetry become completely mixed, symmetry becomes one of the fundamental tools of science, especially of physics and chemistry.

Often you may not realize that you are dealing with an aspect of symmetry. Let's begin with an application. See if you can spot where symmetry enters in.

Consider a game in which a volunteer hides a quarter in one hand and a half-dollar in the other. With a little misdirection, you can get the volunteer to multiply the amount in one hand by an even number and the amount in the other hand by an odd number. From the sum of these two products it is possible to tell which coin is in which hand.

The basic idea behind the method is at least as old as the Pythagoreans. Perhaps the method was invented by old Pythagoras himself, since it is based on one of the differences that he thought was fundamental to the structure of the universe—the difference between even and odd numbers. As subsequent chapters will show, Pythagoras was right about odd and even being fundamental in ways that he could not have dreamed of.

Let's review some of the Pythagorean beliefs about numbers. Pythagoras believed that the world was based on the natural numbers—1, 2, 3, 4, and so forth. The Pythagoreans felt that each of these numbers had special connections with the nonnumerical real world. Since all natural numbers can be formed by successive addition of the number 1, it is the basis of the numerical universe; and since reason alone can tell you how the universe works, 1 is also the number of reason. The number 2 is the first even number; because they believed even numbers were female and odd numbers male, 2 is the number of opinion (the Pythagoreans were also early sexists). As 3 is the first male number (probably reflecting male genitalia), it is the number of harmony; or perhaps it represents harmony because it reconciles the one (1) and the many (2). A square has 4 corners, so 4 represents a square deal, or justice. When you combine male and female, or 3 and 2, you get 5, the Pythagorean number of marriage. The number 6 is the first perfect number—that is a number that can be formed by either addition or multiplication of the same numbers, in this case 1, 2, and 3—so it represents creation. And so on.

Pythagoras thought of the even numbers and odd numbers as female and male. Furthermore, Pythagoreans carefully noted how the types of numbers could be combined.

male + male = female

female + female = female

male + female = male

male × male = male

female × female = female

male × female = female

If the Pythagoreans were a bit sexist, at least the products showed that females dominate four out of six interactions.

Now do you see how to tell which coin is in which hand? Say that the amount in the volunteer's right hand is multiplied by an even number and the amount in the left hand by an odd number. The amount of a quarter, $0.25, is odd, while the amount of a half-dollar, $0.50 is even. You can set up a table. There are only two rows, since if the half-dollar is in the right hand, then the quarter must be in the left, and vice versa. Since the amount in the right hand was multiplied by an even number, the product is even no matter which coin is there. But since the left hand is multiplied by an odd number, the product is even or odd depending on the amount in the hand. Knowing the evenness or oddness of the product for the left hand would suffice, but this might be too obvious. Adding the two products provides another level of mystery, although not any additional information.

Right Hand		Product	Left Hand		Product	SUM
Even	(.50)	Even	Odd	(.25)	Odd	Odd
Odd	(.25)	Even	Even	(.50)	Even	Even

If the sum is odd, it is only possible for the right hand to hold an even amount and the left to hold an odd amount. If the sum is even, the opposite must be the case. The actual amounts do not matter at all; the trick could be performed with a nickel and a dime or with a dime and a penny (but not with a nickel and a penny).

Very nice, but what has this to do with symmetry? The connection will become clearer as you learn more about symmetry in later chapters. At this point, you should notice that some operations always change evenness to oddness and oddness to evenness. For example, the operation of adding 1 will always change evenness to oddness and vice versa. Other operations never change evenness to oddness nor oddness to evenness. For example, adding 2. Mathematicians and physicists say that the first kind of operation changes the oddness or evenness, while the second kind leaves the oddness or evenness unchanged. In mathematics, *parity* is the general term for the same oddness or evenness, so we can talk about operations that change parity or leave it alone. As with *symmetry,* however, the meaning of parity can be expanded to go beyond oddness and evenness. In its expanded forms, parity is more like its base meaning, which refers to any two things that are equal in some sense. Usually in mathematics or physics, parity refers to some quality that exists in two states and that can be changed from one to the other. Any entity in one of the two states has the same parity as any other entity in that state—from the parity point of view they are equal, just as from some points of view all even numbers are the same thing.

Consider the parity difference between opposite and direct transformations that was briefly discussed in Chapter 3. Start with a triangle that has been labeled to indicate a particular sense. The sense of a triangle is the way you travel around it with the labels in order. The triangle *ABC* has a different sense from triangle *ACB.* One reflection reverses the sense. Two reflections reverse it again, returning it to the same sense it had previously. Three reflections reverse it again. This behavior is similar to the effect on parity of adding 1 repeatedly. The operation changes the parity each time, returning the number to its original parity after an even number of operations. You can, then, talk about an odd transformation or an even transformation as meaning the same as the geometer's names, *opposite* and *direct.* Reflection and glide-reflection are odd or opposite; they reverse the sense of the plane. Translation and rotation are even or direct; they keep the sense of the plane the same. If you picture transformations in terms of walking around and around a plane so that you view the triangle from different points on each side of the plane, the parity of the number of times you pass through the plane tells you whether or not the sense will be preserved in each observation of the triangle.

Almost anything that has two different states and an operation that changes one state to the other one can be described in terms of parity. For example, the computer on which I am writing this paragraph operates on that principle. At the operating level, the computer consists of thousands of "switches" that can be either on or off. As I type, each keystroke changes the parity of some of those switches. When I decide to save this paragraph, I will tell the computer (by changing some more parities) to record the paragraph on a floppy disk. The computer will send a signal to the disk drive that will cause it to change the parity of small magnetic domains on the disk in a pattern that reflects the state of the "switches" in the computer.

In mathematics, the idea has also been generalized to more than two states. For example, there is an arithmetic of whole numbers in which all the natural numbers that have the same remainder when divided by a particular number are considered the same. If you divide any natural number by 3 you can only have three possible remainders—0, 1, or 2. In this system, 27 and 36 are considered the same (remainder 0), while 145 (remainder 1) is different from 362 (remainder 2) and the same as 1112 (remainder 1). Thus, when 3 is the divisor, there are three parity states. Similarly, if n is the divisor, there are n parity states.

For most of this book, we will be satisfied with just two parity states. Sometimes they will be even and odd, but often they will be something else.

A Puzzle in Parity

Since people often do not think about the parity of a particular system, the idea has been used for many simple parlor stunts, similar to the trick with the coins. For example, turning a card or a coin over once reverses face and back or heads and tails. If you have a set of cards placed randomly on a table, you can assign 0 to "face down" and 1 to "face up." A quick count of the number of face-up cards will give you a number that is either even or odd, telling you the parity of the system. Then you can ask a person to flip over a number of cards at random while you are out of the room. When the operation has been completed, one of the cards is covered by the person who has done the flipping. On your return, you are told only how many have been flipped, not whether they were flipped from front to back or from back to front. A glance at the table will show you whether the card covered is face up or face down. (Say that the original number of face-up cards was 5, giving an odd parity to the system. Each flip reverses the parity of the system, so if 8 flips are performed, the parity will be unchanged. When you look again at the table, you can count the cards to ascertain the parity that is shown. If it is odd, the hidden card must be face down. If it is even, the hidden card must be face up.)

Another trick based on parity is familiar to most people. It is the 15 puzzle, first developed by Sam Loyd, the puzzle genius of the nineteenth century. In the original form of this puzzle, 15 square tiles were set permanently into a tray that had room for 16 square tiles. That is, the tray provided 15 "cells" for the 15 squares and one "cell" for the "hole." As a result, the squares could be changed around by moving them successively into the square "hole." The tiles were arranged in numerical order from left to right except for the last two, which were interchanged as shown in Figure 4–1.

1	2	3	4
5	6	7	8
9	10	11	12
13	15	14	

FIGURE 4–1
The original form of the 15-puzzle had the squares for 14 and 15 reversed when it was made.

Loyd's challenge to the public who bought the puzzles was to rearrange the squares so that they were all in strict numerical order. He offered a $1000 prize to anyone who could perform the rearrangement and demonstrate how it was done. Some claimed to have solved the puzzle, but no one ever collected the reward.

It is easy to understand why if you begin with a simpler version of the puzzle. Use just three squares with the numbers 1, 2, and 3 along with a square hole. If you begin with 1–3 in the top row and 2-hole in the bottom row, the goal would be to produce 1–2 in the top row and 3-hole in the bottom row (Figure 4–2).

Each time you make a move, the hole can either go back to where it was in the previous move or you can make exactly one new arrangement. You will quickly see, however, that the goal cannot be obtained, for each arrangement leaves (in clockwise order and omitting the position of the hole) 1-3-2 the same. If you start at 1 and proceed clockwise around the square, the numbers stay in 1-3-2 order. To get a reversal, you need to lift two squares out of the puzzle and rearrange them outside the frame. Clearly, in this instance, the two different arrangements (1-2-3 and 1-3-2) have different parity.

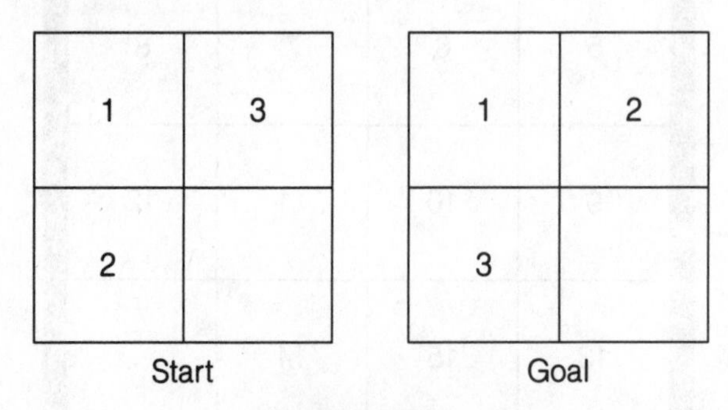

FIGURE 4–2
There are only two moves possible at any given time with the 3-puzzle, but it is easy to check that no combination of moves changes the basic pattern of the numbers. If you start with 1 and proceed around the puzzle clockwise, you will always reach 3 before you reach 2.

If you compare this with what happens to the sense of a triangle under transformations, you will see that the two situations are much the same. Sliding the squares into and out of the hole is a "direct" transformation of the puzzle. To get the sense of the puzzle reversed, you have to leave the plane of the puzzle and perform an "opposite" transformation on it (by lifting squares out of their cells and rearranging them). If the numerals all had line symmetry (e.g., 0, 1, 8), you could also resolve the puzzle by reflection. Try it in a mirror.

The 15 puzzle is the same situation as the 3 puzzle with regard to parity, but more complex because of the larger number of squares. Instead of a dozen different arrangements, you have more than 20 million—but all with the same parity. If you interchange the squares by removing them two at a time from the puzzle and replacing them, instead of simply sliding them around, you reverse the parity (which is clearly what the people who claimed to have solved the puzzle did). Interchanging two of the squares gives another 20 million arrangements with a different parity. You can recognize this as parity because the two states are treated as the same—one can be "solved," and the other cannot.

As with even and odd numbers, we are dealing with two states and an operation that changes one state to the other. If you make an even number of extrapuzzle interchanges to get a desired configuration, the parity will stay the same and the puzzle can be solved. If you make an odd number of extrapuzzle interchanges to get a desired configuration, the parity will change, which means that the puzzle as it was before those interchanges cannot be solved. Since the desired configuration as the original puzzle was sold can be reached by removing 14 and 15 and interchanging them outside the puzzle, which is one extrapuzzle interchange and therefore an odd number of interchanges, the original puzzle cannot be solved.

A methodical way to look at the 3 puzzle (the simplified form of the 15 puzzle) is to watch the way that the hole moves. If you always go to a new position each time, the hole moves around and around the puzzle in the same direction (clockwise or counterclockwise, depending on your first move). The hole travels around the puzzle through three different positions before you get back to where you started. There is no other place for it to go, since there are only four positions in all. It is often more instructive to watch the hole and not the doughnut. Some physical examples that require hole-watching will also occur in succeeding chapters.

Hole-watching can be used to find how to examine a thoroughly scrambled 15-puzzle to see whether or not it can be unscrambled into numerical order with the hole in the lower right-hand corner. First note that the hole, if it starts in the lower right-hand corner and returns to that spot must travel both as many moves to the right as to the left, and as many moves down as up.

Therefore the number of moves for the hole to return to its lower right-hand corner must always be even. If the hole needs an odd number of moves to get to the corner, the position with the hole in the corner could not have been present at the start. In other words, the hole must have been somewhere else that can be reached in an even number of moves. The hole cannot leave someplace and then move back to it in an odd number of moves.

This concept can be extended to describe the state of the entire puzzle. When 15 is before 14 in the puzzle (and all the other numbers are in their proper places, as is the hole), it is a single inversion of two numbers. If 15 were before 14 and 13, there would be 2 inversions. With 2 inversions, the puzzle can be solved. If you count all of the inversions, you can determine the parity state of the puzzle. If the puzzle has an even parity—that is, the sum of all the inversions is even—it can be solved. If it has an odd parity, the solution is impossible.

Parity Squared

Parity is also an essential element in some other recreations. Since the chessboard consists of equal numbers of black and white squares, for example, parity considerations can figure into problems based on chessboards, in which the essential quality is the color upon which a piece sits. It is obvious, for example, that a bishop's moves always maintain the same parity (color), while a knight's moves always reverse the parity of its position. Keeping track of the parity of a given position on the board is an elementary way to keep your pieces from being captured by bishops or knights.

The parity of the chessboard moves for knights is the basis of a chess problem that seems much more complicated than it really is. Use it to baffle your chess-playing friends: What is the maximum number of knights that can be placed on a chessboard so that no knight can attack another? The surprisingly easy answer is 32, since each knight can be placed on one square of the same color as all the rest. None of the knights can be attacked, since each move by a knight causes a change of parity, and there are no knights on the other color.

A number of other puzzles have been proposed based on the parity properties of a chessboard. One of the oldest is the question of whether or not dominoes that are the size of two squares on the board can cover the board when two opposite corners of the chessboard have been removed. No overlaps or hanging over the side allowed is for the dominoes. It helps to look at a chessboard (Figure 4–3).

FIGURE 4–3
Notice that the opposite corners of a chessboard are always the same color.

The opposite corners are always the same color, either both white or both black. If you remove both of them, you do not change the parity situation for the board, which still has the same pattern of black and white squares. However, the twoness of the parities of the domino is not consistent with the overall parity of the altered board. Each domino covers one black and one white square of the board. Therefore, for the domino to have the same total parity as the chessboard, you would have to remove one black and one white square (for example, the two adjacent corners). Removing two squares of the same color, say black, leaves the board with a different parity situation from the domino. With two black squares removed, the board will have an uneven number of pairs of black and white squares. Even though both 32, the number of white squares, and 30, the number of black squares, are even numbers, the total number of pairs of squares is uneven. This is not just a pun on the word *even.* The Pythagoreans used to use dot patterns to represent numbers. An even number is always "even" in such a pattern, while an odd number always has an "odd" dot left over (Figure 4–4).

The same patterns result if each dot represents 2 things or some other number of things as when each dot represents 1 thing. The property of evenness or oddness goes beyond just the even and odd numbers.

An interesting question, then, is whether you can cover the board with dominoes if equal numbers of black and white squares are removed from anywhere on the board. The mathematician Ralph Gomory has shown that it is always possible to remove one square of each color from the board and still cover it with dominoes, no matter where the squares are. If you permit

```
•  •  •        •  •  •
•  •  •        •  •
   6              5
```

FIGURE 4–4

In an even number of dots, the dots can be paired, but an odd number of dots always leaves one left over.

more than a single square of each color to be removed, situations can arise in which a square is isolated or left in a group with odd parity, making the covering impossible.

Another way that the concept of parity extends beyond even and odd involves positive and negative numbers, often called *signed numbers* since they are represented by a positive or negative sign. For any signed number, there are two possible states, positive or negative. Furthermore, there is an operation that changes one state into the other, the operation of multiplying by a negative number. If you remember the operations table you learned in first-year algebra or in junior high, you will recall that

positive times positive = positive
positive times negative = negative
negative times positive = negative

and

negative times negative = positive

The last of these rules is the one that sometimes gives people trouble because it is difficult to picture a physical model for the operation of multiplying a negative by a negative. About the best one that has been developed is to picture a water tank that can be filled (+) or emptied (–). You can make a movie of what is happening to the tank. When the movie is projected, you can run the projector either forwards (+) or backwards (–). A movie of a tank being filled that is running forward will show the tank being filled (+ times + = +), while if the movie is running backward it will show the tank being emptied (+ times – = –). Similarly, if the movie is of the tank being emptied, when the movie is shown forward, the tank will appear to be emptying (– times + = –). But a movie of the tank being emptied that is shown backwards will appear to show the tank being filled (– times – = +).

The simplest way to change the parity of a number (with respect to its sign, rather than with respect to its evenness or oddness) is to multiply it by –1. This operation has been given a name; it is called *taking the opposite* of the number. In fact, taking the opposite is often used as a primitive concept that is the basis of developing the meaning of signed numbers. You can define the operation of taking the opposite in terms of distance without reference to -1. Two different numbers are opposites if they are the same distance from 0. Confusingly, mathematicians use the same sign to mean taking the opposite, subtraction, and a negative number (although it is easy to see that the ideas are related; if x is some number, then $0 - x$ and $-x$ are the same number, although if x is a negative number to begin with, $0 - x$ and $-x$ represent a positive number). Another way of thinking about the sign $-$ is just to say that it reverses a number's parity. Thus, $-x$ always has a different parity from x, whether x is positive or negative—with one important exception. If x is 0, the number $-x$ is the same as x. Therefore, 0 is its own opposite, so for $x = 0$, the parity of x and $-x$ is the same. In practice, this turns out to be much less confusing than it seems at first.

Parity based on evenness and oddness and parity based on positive and negative are more deeply related than you would expect. To show this, you need to examine some graphs. Figure 4–5 shows the graphs of two simple algebraic equations.

The first way we will look at these two graphs is by reflecting each of them in the vertical, or y, axis in the plane in which

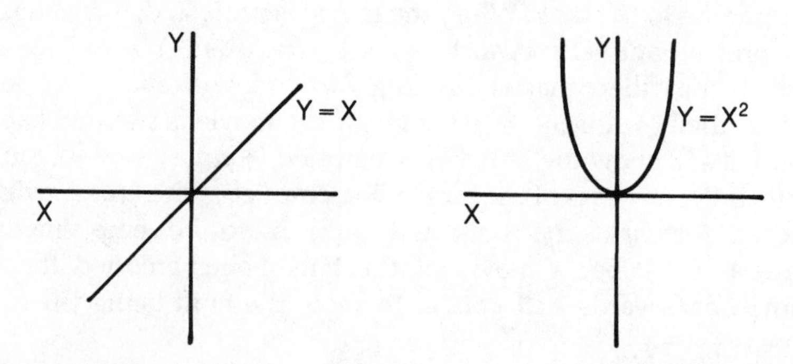

FIGURE 4–5
The two graphs are different with respect to the kind of transformations that will leave the curve unchanged.

each is graphed. What happens? The graph of $y = x$ is reflected into an entirely different line, but the graph of $y = x^2$ is reflected into itself. Notice, however, if you choose a different transformation, another behavior will be observed. Reflect each graph through the point where the two axes cross (called the *origin* of the plane). In this case, the graph of $y = x$ will transform into itself, while the graph of $y = x^2$ will become a different graph.

Now look at the graphs from the point of view of parity with respect to sign (Figure 4–6). Change the parity of x in each equation by substituting its opposite, $-x$. The left-hand graph becomes the graph of $y = -x$, while the right-hand graph becomes the graph of $y = (-x)^2$, which is the same as the graph of $y = x^2$.

How does this compare with what happened when you reflected each graph about the y axis and through the origin? For these two graphs, at least, the result of reflecting about the y axis is the same as the result of substituting the opposite for x. A little thought should suggest to you that this will always be the case, for if (x, y) is a point on a graph, then $(-x, y)$ will always be the reflection of that point in the y axis. On the other hand, the graph of $y = x^2$ did not change when $-x$ was substituted for x because the points in the reflection are all already on the graph. When you reflect a point through the origin, however, (x, y)

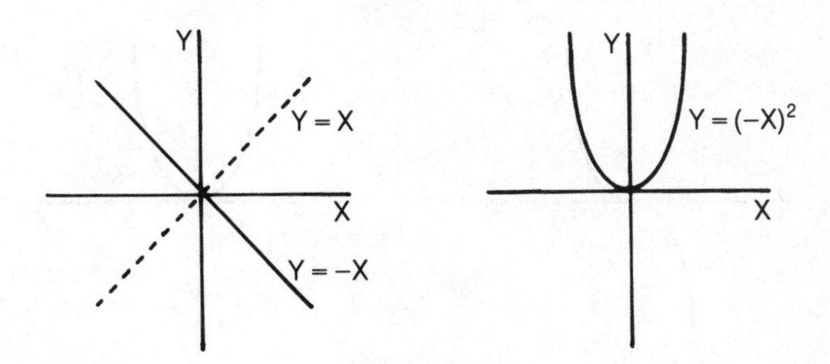

FIGURE 4–6
Substituting $-x$ for x in $y = x$ produces a different graph ($y = -x$), but substituting $-x$ for x in $y = x^2$ does not change the graph.

becomes $(-x, -y)$. Thus, the graph of $y = x$ becomes the graph of $-y = -x$, which is the same equation (after a little fiddling with it) as $y = x$. On the other hand, the graph of $y = x^2$ is changed by reflection through the origin into the graph of $-y = x^2$, an entirely different equation no matter how you fiddle with it.

Look at Figure 4–7. The graph of $y = x^3$ is like the graph of $y = x$ as far as its behavior when reflected through the origin goes. That is, it does not change. Similarly, the graph of $y = x^4$ is like the graph of $y = x^2$. It does not change when reflected through the vertical axis. Further investigation would show that all the graphs of $y = x^n$ have the following property:

- If n is even, the graph does not change when reflected in the vertical axis.
- If n is odd, the graph does not change when reflected through the origin.

In $y = x$, of course, the exponent is 1, so the graph obeys the rule for odd exponents.

There is another way to look at this interesting relationship between equations and the kind of symmetry their graphs possess. You know that reflection through the vertical axis means replacing the point (x, y) with the point $(-x, y)$. Therefore,

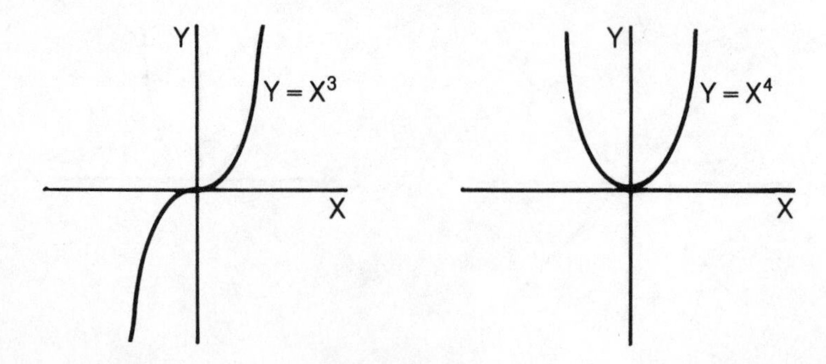

FIGURE 4–7
Note that the graph of $y = x^3$ is like the graph of $y = x$ in that reflection in the origin does not change either of them, while the graph of $y = x^4$ is like the graph of $y = x^2$.

another way to state the relationship is as follows: An equation has a graph symmetric about the vertical axes if and only if the value of y is unchanged when x is replaced in the equation by $-x$. Similarly, reflection through the origin means that (x, y) is replaced everywhere by $(-x, -y)$. Therefore, the graph will be symmetric about the origin if and only if the value of y is changed to $-y$ when x is replaced by $-x$ in the equation. These relationships will hold true for more complex equations than those of the simple form $y = x^n$.

It is convenient at this point to translate what we have been discussing into a different mathematical representation, based on the concept of a *function*. A function is any rule that associates two sets of numbers in such a way that if you know one of the numbers in the first set, then you also know the number in the second set with which it is associated. In practice, for the functions used in algebra, both sets are taken to be the same set—the set of real numbers. Real numbers are any numbers that can be represented by decimals, such as 3, −5, 63.7, $1/3$, and $\sqrt{2}$. The nature of the function may restrict which real numbers can belong to one of the sets; for example, the function defined by the equation $y = x^2$ has all the real numbers associated with the first set (which is conventionally assigned to x), but the numbers in the second set do not include any negative numbers, since no real number has a square that is negative.

When using functions, special notation is employed. The most common notation denotes the value of a member of the first set as x and the value of the corresponding member of the second set as $f(x)$. In other words, $f(x)$ replaces y in the kind of equations we have been discussing. The equation $y = x$ can be written in function notation as $f(x) = x$. In this form, you can refer to the whole relationship as f. That is, f is the name of the function defined by $f(x) = x$. If more than one function is being discussed, other letters can be used; by convention, the most commonly used other letters are F and g, although any letter or other symbol can be used.

Now to get to the point of this digression: In function notation, the rule for determining whether a function is symmetric about the vertical axis or through the origin can be stated simply.

If $f(-x) = f(x)$, the graph (and by extension, the function) is symmetric with respect to the vertical axis. If, on the other hand, $f(-x) = -f(x)$, the graph and the function are symmetric with respect to the origin. Following the first observation about functions of the form $f(x) = x^n$, functions for which substituting the negative of x for x produces the original function are called *even,* and functions for which substituting the negative of x for x produces the opposite of the original function are called *odd.*

By no means are all functions either even or odd, which may seem strange since all counting numbers that are not even must be odd. Functions behave more like real numbers. Is $1/3$ odd? Is $\sqrt{2}$ even? These questions have no meaning, even though an infinite number of real numbers are either even or odd.

Knowing the parity of a function is a great help in sketching its graph. It cuts the work in half if the function is either even or odd, because you can sketch the other half as the appropriate reflection of the half you have already worked out.

An Algebra of Geometry

If you combine two transformations, the result is another transformation. The way that the combination is expressed in mathematical symbolism can be either as a sum or as a product, depending upon the particular inclination of an author and the habits in that branch of mathematics. Of course, the combination is neither a sum nor a product in the ordinary sense of addition or multiplication of numbers. Sometimes confusion is avoided by using some operation sign other than + or ×. For example, you can use *, to mean the combination. In this form, the sign would be read as either "followed by" or "preceded by." In this section, a compromise from this welter of conflicting symbolism will be used. The symbol * will be used, and it will be read as "preceded by," but the result of the operation will be called the *product* of the two transformations.

In Chapter 2, a translation was defined as the result (now you can say *product*) of two reflections in parallel lines. A translation can also be defined as the product of two half-turns about two points, or, equivalently, two reflections in two points. If you agree that the identity transformation is also a translation (one that does not translate the figure very far, of course), then the product of two reflections about the same line or of two half-turns about the same point is also a translation. If you use the letter R for reflections in a line, H for half-turns about a point, and T for translations, you can write

$$H_2 * H_1 = T$$

for any two half-turns about different points. If the reflections are in parallel lines, you can also write

$$R_2 * R_1 = T$$

The first of these means that the half-turn H_2 preceded by the half-turn H_1 is a translation. It is a custom that the transformation on the right is done first, followed by the transformation on the left. You can now explain the second equation in terms of its meaning as well.

The next question is how do the other transformations you have been looking at combine? For example, what is $H * T$ or $T * H$? You can work these out by visualizing them or by making a few sketches. Note that

$$H_1 * T = H_2$$

and

$$T * H_1 = H_2$$

where H_1 and H_2 are just half turns about two different points (unless T is the identity translation).

What is the product of two translations? The result is just another translation. Similarly, what is the product of two reflections through different lines when the lines are not parallel. You already know the answer to this. When the lines are not parallel, they must intersect in a point. The definition of rotation about a point was originally framed in terms of this situation. Since R is reserved for reflection, you can use S for rotation:

$$R_2 * R_1 = S$$

Now consider the product of two rotations. Most of the time, the result will be another rotation (but about some different point). If the sum of the rotation is 360°, however, the effect is going to be the same as two half-turns; in that case, the product will be a translation, as before.

The product of a rotation and a translation is a rotation, also. It will be a rotation through the same angle, but with a different center.

Work out for yourself the combinations that yield glide-reflections and the combinations of glide-reflections.

All of these relationships form a kind of algebra that can be used to explore geometry. You have already seen examples of the basic idea in Chapter 3, where the transformations were not quite so abstract. Entire books have been written that develop all of geometry from a transformational point of view.

Eschervescence (Part 2)

In many of his works, M. C. Escher transforms a figure in two different ways, combining a symmetry transformation such as translation with a *color transformation.* Until recently, the color transformations were not considered part of mathematics or science. After all, no one cares whether a triangle is in black ink on white paper or in white ink on black paper. But today it is clear that color transformations form an important branch of symmetry theory that also has applications in crystallography. The color change is nothing more than our friend parity in another guise. That is, if you change the color from black to white an even number of times, the color remains the same, while if you change an odd number of times, the color changes. Figure 4–8 is a simple example in which the parity transformation is combined with translation.

The reason that such parity transformations were of interest to science is that crystallographers realized that certain crystals showed properties that are similar to the parity transformations.

FIGURE 4–8
The combination of translation and changing color is used in this work by M. C. Escher.

For example, crystals of potassium chloride when examined by x-ray diffraction techniques show both the potassium ion and the chlorine ions as exactly alike. It is as if the x-rays were "color blind" to the differences. When crystallographers describe these crystals, they find it mathematically more useful to say that the symmetry property is a combination of the symmetry of a cube (since the crystals are cubical) and the parity transformation.

Color symmetry was first suggested to crystallographers in 1930, but it was not until the 1950s that it became an important concept.

All sorts of transformations can be combined with the parity transformation to produce a figure that represents a different symmetry. Escher has made the most of this in his many drawings and woodcuts. The works with parity transformations are among his best known. The flying horses once flew along the cover of a book on crystal symmetries, while the birds in Figure 4–9 flew on the cover of *Scientific American.*

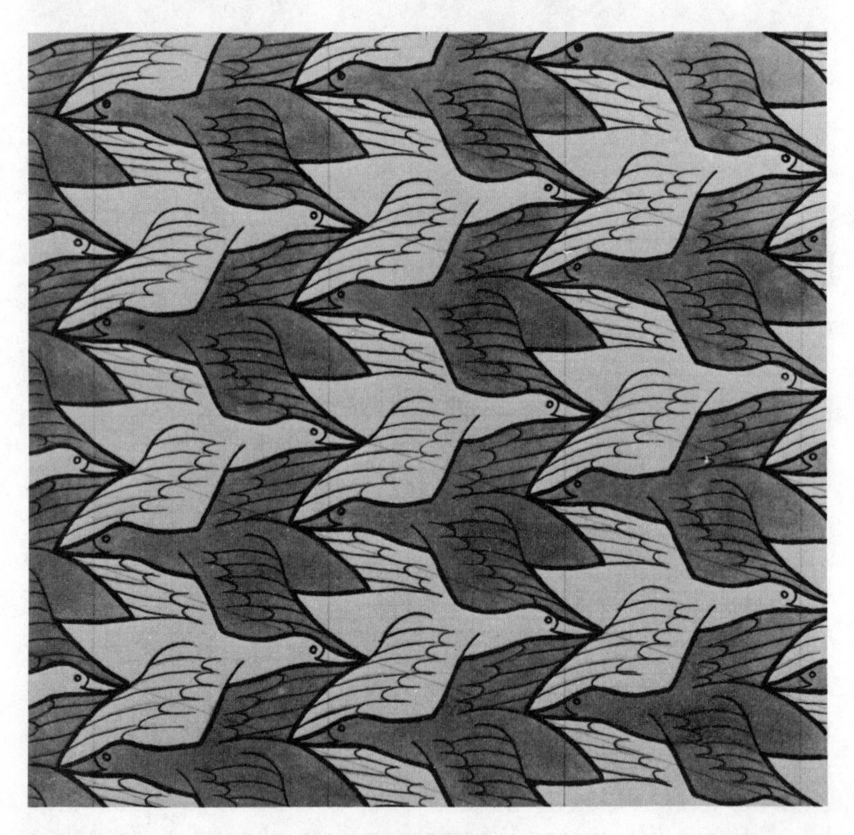

FIGURE 4–9
At first glance, Escher's black and white birds appear to be a combination of glide-reflection and a color transformation, but look more closely. © 1989 M. S. Escher Heirs/Cordon Art—Baarn, Holland

The flying birds are not, as a first glance would suggest, an example of a parity transformation combined with a glide-reflection. If you look more closely, you will see that the tails of the white birds are raised, while the tails of the black birds are lowered. As far as I know, there is no mathematical description of raising and lowering tails as a transformation, but no doubt birds know all about it. In any case, it is not a symmetry transformation.

A parity transformation always has to be combined with some other transformation for the result to be shown. The object cannot be black and white (or potassium and chlorine) at the same time. It is easy to imagine, however, a parity transformation combined with a translation along the time axis. Such transformations occur in nature with some regularity. For example, the arctic ptarmigan is a bird that is white in the winter and brown in the summer. You are dealing with the same bird, but it has been translated 6 months in time and gone through one parity transformation. If you wait any even number of half-years, the bird will look the same. But if you wait an odd number of half-years, the bird's color will be changed.

One Kind of Parity for Particles

The even and odd functions are familiar to mathematicians, but most of them would be surprised to discover that there is a striking analogue to these functions in particle physics—that part of physics that deals with the properties of electrons, protons, and neutrons as well as the roughly two hundred other particles that are to one degree or another exotic. Particles, like functions, have been classified as even and odd. Furthermore, with suitable definitions for the meaning of the terms, if $p(x)$ is a particular particle, and if $p(-x) = p(x)$, then the particle is classified as *even.* Also, if $q(x)$ is some other particle with the property that $q(-x) = -q(x)$, then the particle is classified as *odd.*

It is not clear without some definitions, however, as to what is meant by the description in the last paragraph. One way to look at this kind of parity of particles is to picture the particle as being reflected in a mirror. That is, using $p(x)$ as the original particle, $p(-x)$ is the reflection. Say that the original particle is at a position in space identified by the three coordinates (x, y, z). Then the reflection in the origin has the coordinates $(-x, -y, -z)$, as you know from your experience with transformations. Now, given that change in sign of the coordinates, the reflection can be either exactly the same as the original particle, or it can have left and right reversed. Left and right for particles can be described in terms of a particle's *spin*. All particles have a definite spin. Spin is measured as angular momentum, similar to the numerical measure that combines the speed of revolution and mass of a top. The angular momentum of each spin is a multiple of a single number, which is about 1.0544×10^{-27} g-cm^2/s (this number is Planck's constant, *h*, divided by π). The angular momentum of an electron in orbit about the nucleus of an atom is exactly twice this number, or some multiple of twice the number. The angular momentum of an electron not in orbit is $1/2$ the angular momentum it has in orbit. Therefore, the spin of the isolated electron is referred to as $1/2$. For practical purposes, physicists always refer to the spin as if it were a number without units. Therefore, all particles have a spin that is expressed as a multiple of $1/2$.

If the spin of a particle is an integer (0, 1, 2, . . . are all multiples of $1/2$, of course), mirror reflection will not change the direction of spin. Therefore, particles with integer spins are even. Furthermore, particles with nonintegral spins are all odd. Electrons, protons, and neutrons are among the particles with an odd spin. Photons and some mesons are particles with an even spin.

Although this system of classification is a typical parity system, it is *not* what physicists today mean when they speak of parity (although some similar ideas are central to both the type of parity discussed here and the type that is usually meant by the word *parity.*) Recall that there is a second meaning of parity, which was called "plus-or-minus parity." It is the plus-or-minus parity of particles that is generally meant when physicists talk about parity. Specifically, the law of conservation of parity refers to the plus-or-minus classification. This aspect of symmetry will be discussed in detail in Chapter 7.

The implications of the even-odd parity for particles are much greater than you would expect. It is easier to look at the one-dimensional case than the three-dimensional one; however, the results are the same. In one dimension, you are dealing with a pair of particles, the original particle and its mirror image, at positions that can be identified as x and $-x$. These two particles are completely indistinguishable, except possibly for the direction of their spins. If you interchange the two particles, the result for the system of two particles can be one of two things. Either the system will be exactly the same, in the case where the particles are even; or it will be the reverse of what it was, in the case where the particles are odd.

The same would hold true for two actual particles. If the two particles are indistinguishable, interchanging a pair of even particles will produce no change, while interchanging a pair of odd particles will reverse the state of the system.

It is useful sometimes to treat particles as waves. At subatomic levels, particles are waves and waves are particles—an idea that seems paradoxical but that has been confirmed by countless experiments. The result from a particular experiment may be better interpreted by viewing the fundamental truth behind the experiment in terms of particles or in terms of waves. Both are only interpretations of the real fundamental truth, a truth that seems to be beyond our understanding.

The wave interpretation is the more difficult one to comprehend, partly because the waves are waves of probability, which is a far cry from the kind of waves you see at the beach. It is more complicated than that, for the probability is not the wave itself, but the square of the wave.

One useful interpretation of probability waves returns to the particle idea. The exact position of a particle in space cannot be predicted before an experiment is performed. The square of the particle's wave function describes the probability that the wave is at some particular point in space.

Now apply that to the case of a particle and its mirror image. Since the two particles are indistinguishable, you cannot tell whether or not they have been interchanged. If $P(x)$ is the probability that one of the particles is at point x, then $P(-x)$ is the probability that the particle is at point $-x$. Necessarily, $P(x) = P(-x)$, for you must not be able to tell which particle is where since $P(x)$ and $P(-x)$ are completely indistinguishable. The square root of the probability is the wave function. But there are two square roots, one positive and one negative. If the two positive or the two negative square roots are taken, you get an even wave function. If a positive square root is paired with a negative square root, the wave function is odd. Since these are the only two possibilities, all wave functions must be either even or odd (unlike functions in general, which can be neither).

The graph of an even wave function for two indistinguishable particles, then, will look something like the sketch in Figure 4–10.

Similarly, the graph of an odd wave function for two indistinguishable particles will look something like Figure 4–11. Notice that the graph must cross the origin to be odd.

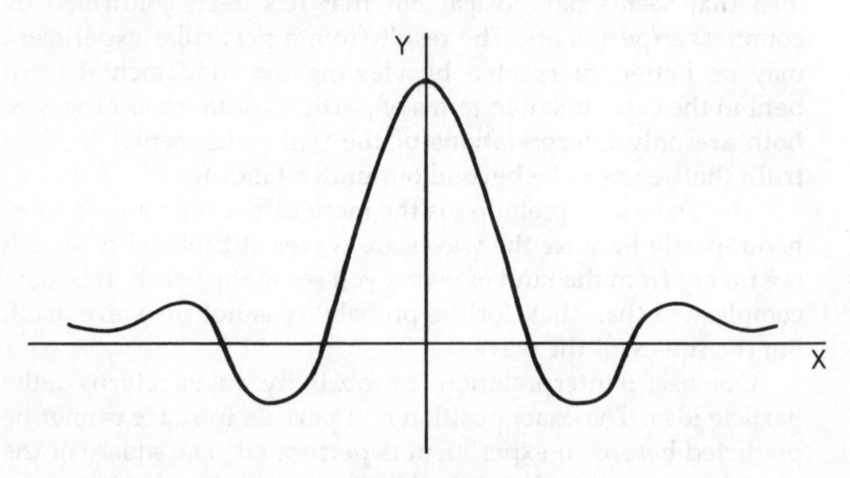

FIGURE 4–10
The graph of a wave function of a particle can be even.

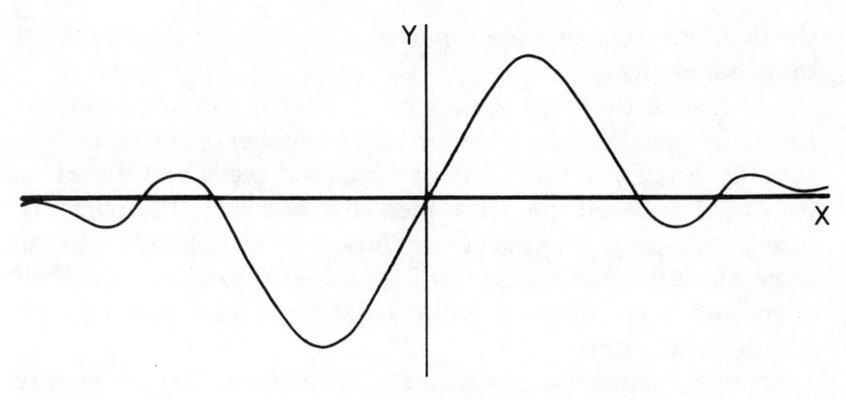

FIGURE 4–11
The graph of a wave function of a particle can also be odd. In fact, all particles have wave functions that have either even or odd graphs.

This result has important physical implications. Picture the two particles in the system as coming closer together, so that they approach the vertical axis. Since the probability of position is the square of the function, for an odd function the probability of the particles being in the same place (at the origin) is 0. Since the graph crosses the axis at the origin, the square of the value of the graph on the vertical axis will be 0 also. Similarly, as two even particles approach each other at the vertical axis, the square of the function will always be positive (even if the function has a negative value at that point). The position of the vertical axis is arbitrary, that is, it was chosen to be halfway between the two particles. Therefore, the results that follow will be true for any two particles that approach each other anywhere in space.

Look at the even particles first. If they are close to each other, then there is a positive probability that they will be in the same position. A result is that there is an fictitious force that brings the particles together. (Fictitious forces are discussed in more detail in Chapter 6.) Does this actually happen? Yes. After this fictitious force was predicted, experimental physicists found that if a gas of identical even particles is cooled enough so that the intrinsic motion of the particles was slowed down almost to zero (cooling simply means slowing down the average motion of particles), the fictitious force became apparent. The particles all tried to get in the same place.

As I write, the most exciting topic in science is cold fusion, or fusion in a jar. While it is not yet clear whether or not this effect actually occurs, the most interesting (and plausible) theory to explain it is based upon the idea that deutrons, the nuclei of heavy water, are even particles. Thus, they want to all be in the same place. In cold fusion—if it exists—the deutrons get their wish, and when they reach the same place, they fuse to form helium or tritrium.

Now consider the case of the odd particles. The probability that they will be in the same spot is 0. In effect, this means that two indistinguishable odd particles cannot ever occupy the same position in space. A particle can be described by a set of numbers that are called the *quantum numbers* of the particle. The quantum numbers completely describe the particle. In other words, two particles with exactly the same quantum numbers are indistinguishable. As a result, two odd particles with the same quantum numbers cannot ever occupy the same position. This rule, which was discovered by Wolfgang Pauli (1900–1958), is known as the Pauli Exclusion Principle.

Pauli was a very interesting man. His earliest fame came at the age of 21 from an article he wrote about relativity theory for an encyclopedia. The article was so good that Einstein, upon reading it, felt that Pauli knew more about relativity than Einstein did. Pauli was purely a theoretical physicist. Indeed, he was so bad at laboratory physics that an effect was named after him, the Pauli Effect. The Pauli Effect occurred when Pauli visited a physics laboratory; all the equipment began to malfunction. It is said that an explosion in a physics laboratory in Berlin occurred at the exact moment that a train carrying Pauli from one city to another paused to change engines in Berlin. Despite this, Pauli was a brilliant theoretical physicist. He discovered the exclusion principle for electrons (which are odd particles) in 1925 without the benefit of the symmetry principle that explains it, which is even more remarkable.

One of the clues that Pauli used in finding the exclusion

principle was the Periodic Table of the elements. By that time, it was already known that the Periodic Table results from the fact that electrons orbiting an atom form various shells. Each shell holds at most only a certain number of electrons. The innermost holds only 2, the next one out can hold 8, the next one 18, and the next one 32. It was a natural goal of physicists to explain this pattern from first principles. The assumption is that such a pattern does not result from chance, but is caused by something. There is still a lot of Pythagoreanism in science. Pauli was able to establish that the different quantum numbers that could be assigned to electrons in each shell included all the possibilities. That is, there were only two electrons in the first shell because if there were a third, it would have exactly the same set of quantum numbers as one of the two already there. The same principle accounts for the next shell, except that there are 8 possible assignments of quantum numbers. The Pauli Exclusion Principle accounts for all of the shells of the atom, which in turn explains the Periodic Table of the elements, which in turn explains the chemical behavior of matter.

Because of the way that the Exclusion Principle functions in molecules, the shape of a molecule is determined. For all compounds, the shape of the molecule is one of the characteristics that determines the properties and behavior of the compound. For large organic molecules, such as proteins or RNA, the shape often determines the biological activity of the molecule. It would be hard to imagine life without the specific key-and-lock arrangements of these large organic molecules. Discovery of the exclusion principle was a remarkable achievement.

Now, as you have seen, the basis for the Pauli Exclusion Principle is the symmetry of the wave function of the electron. Furthermore, the same symmetry applies to the proton and neutron, which are also odd particles, so the exclusion principle also accounts for the structure of the nucleus. It would be only a slight exaggeration to say that symmetry accounts for all the observable behavior of the material world.

5

Finding the Way

Nature seems to take advantage of the simple mathematical representation of the symmetry laws. When one pauses to consider the elegance and the beautiful perfection of the mathematical reasoning involved and contrast it with the complex and far-reaching physical consequences, a deep sense of respect for the power of the symmetry laws never fails to develop.

C. N. Yang

"Cheshire-Puss," she began, rather timidly . . . "Would you tell me please, which way I ought to go from here?"

"That depends a good deal on where you want to get to," said the Cat.

"I don't much care where —" said Alice. "Then it doesn't matter which way you go," said the Cat.

Lewis Carroll

Although it is still startling to the nonscientist, physicists and mathematicians have become accustomed to the fact that we live in a world of four observable dimensions, not three. (A few years ago, I would not have felt compelled to include the qualifying *observable,* but—as you will see—there are good reasons to do so.) Until 1908, it was generally believed by all that the dimensions of length, breadth, and depth sufficed to describe the real world, although mathematicians had found it convenient for more than 50 years to work with any number of dimensions. What happened in 1908 was that Herman Minkowski reformulated Einstein's new (1905) theory of special relativity in terms of a geometry of four dimensions, with time as the fourth dimension.

At first, no one but Minkowski was impressed, although all who heard him must have been surprised when he stated "Henceforth space by itself, and time by itself, are doomed to fade away into mere shadows and only a kind of union of the two will preserve an independent reality." Even Einstein thought Minkowski's concept was "superfluous learnedness." Gradually, however, Einstein and other physicists generally came to appreciate that Minkowski's geometry was the simplest way to describe reality.

It is an—usually unstated—axiom of physics that the simplest description of the real world that works is accepted as the truth, so the four-dimensional description (until replaced by an even simpler description) is not treated as a mathematical conceit, but is, *in fact*, reality. This principle is actually an extension of the concept of Occam's Razor: *One must not multiply entities*, the common statement of the philosophical principle that the simplest explanation should be accepted. The extension that many physicists make of this principle, however, is that the simplest explanation *is* reality.

This concept of reality can be contrasted with various other mathematical constructs used in physics. A particular description may be chosen because it simplifies the mathematics but is not thought of as being real. The notion of dimension is one of those constructs used to simplify various problems in the real world, as well as being a basic reality itself. But when dimension is used only for problem solving, it has a different face than when it is treated as something real. For example, one way to work out the motions of a tossed pair of pliers uses a 6-dimensional space, but that is not viewed by physicists as a real space. The 4-(or more) dimensional space-time we live in is viewed as real.

To understand how these various ideas about dimension differ, it is vital to begin with a clear understanding of what is meant by *dimension*.

Directions for the Real World

René Descartes liked to lie in bed until 11 in the morning. He claimed that his health required this kind of a regime, but also he did much of his work in bed. He may have been right about his health; when he was acquired as a court philosopher by Queen Christina of Sweden, he was required to get up at 5 A.M. to philosophize for the queen—and was dead within months from pneumonia.

One day—not night!—Descartes was lying in bed watching a fly crawling on the ceiling. He became interested in how the fly's position in the room could be described. He realized that two numbers were required, since the fly's position was uniquely determined by two intersecting lines. In turn, each of the lines could be described by a single number, representing the distance of the line from one of the two perpendicular walls of the room.

This insight does not seem striking today, and, in fact, Pierre de Fermat reached almost exactly the same conclusion around the same time as Descartes (but did not publish his work). It was, however, a new insight at the time and had been overlooked throughout thousands of years of mathematics. Descartes' method of locating the fly in his room went on to become the essential development in applied mathematics. Although both geometry and algebra had advanced considerably as separate disciplines, Descartes' new method combined the two, which led to calculus and nearly all of modern mathematics. Furthermore, it is hard to see how great patches of physics could have arisen without the tools of the combined algebra-geometry.

In Descartes' first application of the insight, he converted the idea from a fly on the ceiling to the geometric plane. In the plane, any point can be specified by its distance from *two* intersecting lines. Although today we normally take these lines to be perpendicular, this is not required (and Descartes did not use perpendicular lines himself). The key notion is that two numbers are needed to specify the location of a point in a plane. These numbers are the two distances from the intersecting lines. By letting the numbers be both positive and negative (unlike ordinary distance, which is always positive), the four different regions of the plane marked by two intersecting lines can be reached with just the two numbers. Most people are familiar with this idea from high school.

Next, you should note that *three* numbers are needed in ordinary space. These are the three directed (that is, positive or negative) distances from three intersecting planes. Therefore, a useful, but ultimately naive, concept of dimension is that a dimension is the number of distances needed to locate a point. Let's explore this a little further.

Mathematicians soon found various clever ways to locate points that did not involve intersecting straight lines or planes. These were generalizations of Descartes' method.

One way that is commonly used to locate points in the plane is to combine the distance from a point with an angle. For example, point P in Figure 5–1 can be located either in the ordinary Cartesian (that is, Descartesian—but the *Des* is *always* dropped) way or in this other way, which is called *polar.* Using the Cartesian method, P is 1 unit from the x axis and 1 unit from the y axis. In the polar method, however, P is $\sqrt{2}$ units from the origin (the intersection of the x and y axes) and 45° from the positive x axis. The Pythagorean theorem can be used to show that these two methods locate P in the same place. The polar method can also be used to locate any point in a plane. Once again, it needs just two numbers.

Both the Cartesian and polar methods can also be thought of as families of curves. In the Cartesian system, the two families of curves are two different families of straight lines—mathematicians call all one-dimensional paths, including straight lines,

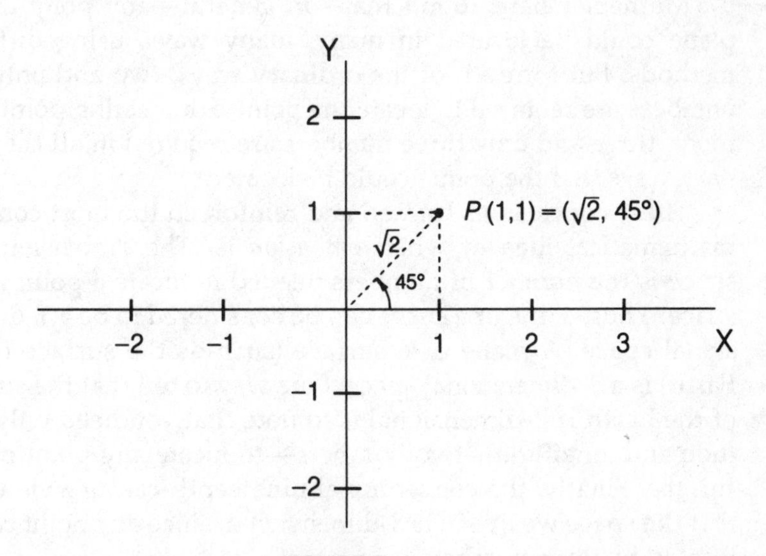

FIGURE 5–1
You can locate a point either in terms of its distance from two intersecting lines (Cartesian coordinates) or in terms of its distance from a point and the angle between two lines (polar coordinates).

curves. The members of each family of curves have a family resemblance; in the case of straight lines, the family resemblance is that the lines are all parallel or perpendicular to each other. In the polar system, one family of curves consists of concentric circles, while the other family is the set of lines through the center of the circles.

In practice, however, this description of the polar system allows two different points to be located by the same two numbers, since a line though the center must intersect a circle in two points. Conventions about signed numbers, however, prevent a given pair of curves from locating more than a single point. For polar coordinates, while distances and angles may be either negative or positive, a given positive or negative distance and a given positive or negative angle combines to locate exactly one point. (The reverse is not true, however; there is more than one way to combine a directed distance with a directed angle to locate a given point.)

Mathematicians found that—in general—any point on the plane could be located in many, many ways, using different methods. But for each of the ordinary ways, two and only two numbers are required to locate the point. For locating points in a room, three and only three numbers are required in all the ordinary ways that the points could be located.

These conclusions formed and reinforced the most common mathematical idea of what *dimension* is. The dimension of a space is the number of numbers needed to locate a point in the space. Thus, a line or a curve can be considered to be a 1-dimensional space. A plane or a surface (such as the surface of the Earth) is a 2-dimensional space. One way to tell that the surface of the Earth is 2-dimensional is to note that you need only latitude and longitude—two numbers—to locate any point on the surface. Finally, the conventional nineteenth-century view was that the space we live in is 3-dimensional, since any point can be located by three numbers.

A naive way to picture the twentieth-century view that space is 4-dimensional is to say that all the points you might wish to locate can be in different places at different times. Therefore, you need to add a fourth number to describe the location of a point. While this naive concept probably helps in feeling comfortable with a 4-dimensional world, it is much less meaningful than the mathematical concept of 4-dimensional space-time, the development to which Minkowski referred. Let me save mathematical spacetime for a few pages while we look at another mathematical construct.

An Interesting Sidelight

The following sidelight may deepen your understanding of dimension. It is not directly required for what follows, but it may add some insights for you. There can be dimensions of ordinary space that are not whole numbers, but which are instead fractions. Fractional dimensions arise from a procedure that is a form of symmetry.

The word *dimension* has two common meanings. The mathematical use of dimension to mean one of the directions in a space (or one of the numbers locating a point) grew out of the use of dimension to indicate a measurement of length. The question "What are its dimensions?" asked about a breadbox would normally be answered "about two feet by one foot by one foot," not, as the mathematician might answer, "three" (nor, as the physicist might answer, "four"). Thus, you make a scale drawing by reducing the dimensions of the object being drawn. You interpret a map by increasing the dimensions shown on the map. In neither case, however, do you mean that you are changing the number of dimensions in the mathematical sense.

To avoid confusion in what follows, we will talk about changing the *scale* and restrict the word *dimension* to its mathematical meaning. If the scale is changed, that is all that is changed. Everything else is invariant.

Figure 5–2 is an example of the consequences of changing scale. A 2-meter by 3-meter rectangle might be mapped with a scale of 1 centimeter to 1 meter.

If the scale were changed to 1 meter equals 1 millimeter, however, the rectangle would have the same shape, but be 1/10 the size (Figure 5–3).

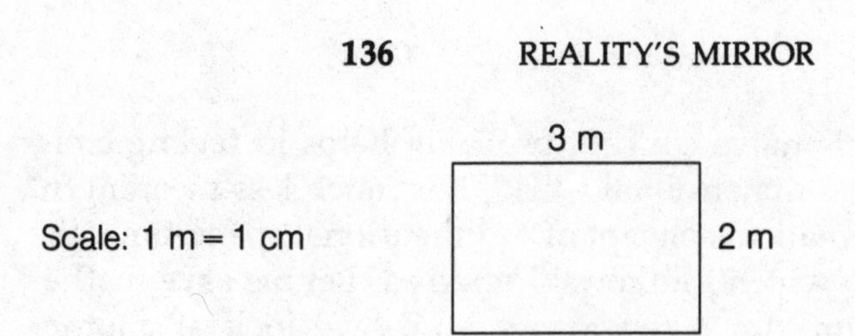

FIGURE 5–2
A 2-meter by 3-meter rectangle when mapped at scale of 1 m = 1 cm.

3 m

Scale: 1 m = 1 mm 2 m

FIGURE 5–3
A 2-meter by 3-meter rectangle when mapped at scale of 1 m = 1 mm.

Consider what kinds of figures for which you could change the scale without changing the appearance of the figure at all. One such figure is the line (meaning, as always in mathematics, a straight line). This line has no thickness and is infinitely long. If you were to look at it in a magnifying glass that increased its scale by a factor of 100, or if you were to view it though the wrong end of a telescope, decreasing its scale by a factor of 100, there would also be no change in appearance. Since the figure is left invariant by a change in scale, this is a form of similarity. Generally, it is called *self-similarity,* since a part of the figure is similar to the whole figure.

Are there other figures in the plane besides the line that are self-similar with regard to changes in scale? It can be shown that in the plane only the line is self-similar for all changes in scale (Leibniz once suggested using this property as the definition of a line), so the answer with this amount of generality is "no." But there might be figures that are self-similar for some specific changes in scale. In other words, the question would become: Are there figures for which there exists at least one number n different from 1 such that n times all lengths in the figure (with other features invariant) leaves the figure as a whole invariant?

The answer is "yes," but it should be noted that many of these curves involve an infinite number of steps to construct and have many counterintuitive properties. When curves with these odd properties were first discovered (by Karl Weierstrass in 1872), mathematicians were reluctant to publish descriptions of the curves. First privately and then publicly, after word of the Weierstrass curve reached print (by another mathematician) in 1875, the curves were denounced as "pathological" and "monsters." Many mathematicians still call them *pathological curves* today, as if that were a name like *simple closed curves.* Benoit Mandelbrot (about whom more will be said later) has proposed a name based on the "monster" designation—*teragons.* Not all teragons are self-similar, but they have such properties as the impossibility of constructing a tangent to the curve at any point or an infinite length that bounds a finite area. As an example, Figure 5–4 is the "snowflake" (also known as the Koch curve or the von Koch curve, since it was discovered in 1904 by Helge von Koch).

The Koch snowflake is formed by successive application of a simple rule repeated an infinite number of times. In this case, a series of equilateral triangles is erected on the sides of an original triangle. In the first step, three new triangles, each with sides one third that of the original triangle, are erected on the edges. In the second step, two triangles are erected on each of the "points"

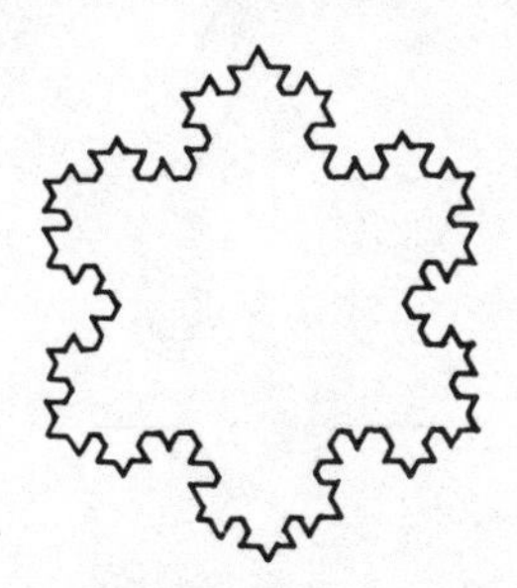

FIGURE 5–4
The Koch snowflake has an infinite length, since it is formed by a process that continually adds "kinks" to the curve, but it has a finite area.

formed in the previous step, for a total of twelve new triangles, forming eighteen new points. For the third step, two triangles can be erected on the sides of each of the new points, forming 18×2, or 36, new points. Figure 5–5 shows how to make a Koch snowflake.

Then, you just repeat the process an infinite number of times.

Originally, the Koch snowflake was famous for a property that was not recognized as being related to dimension. Consider the length of the curve as you pass through the succeeding steps in its formation. Assume that the original triangle has sides 1 unit long, for a total border of 3 units. In the first step of the construction, the new triangles will have sides $^1/_3$ unit long, but there will be 12 sides in all, for a total border of 4 units. The second step produces all straight line segments of the border that are $^1/_9$ unit long, but there are 48 of them, for a total border of $^{48}/_9$ units, or $5\,^1/_3$ units. The ratio of 4 to 3 is the same as the ratio of $5\,^1/_3$ to 4 (check it out). It is easy to show that the same ratio will continue to hold throughout all the steps in the construction of the curve. Therefore, the curve gets longer and longer, becoming infinite after an infinite number of steps. At the same time, the region of the plane that it encloses cannot become infinite, since it remains within a square whose sides are three-fourths of $\sqrt{3}$ units in length. Thus, the area enclosed must be less than the area of that square, which is $1\,^{11}/_{16}$ square units. Even tighter limits on the total area can be given by looking at

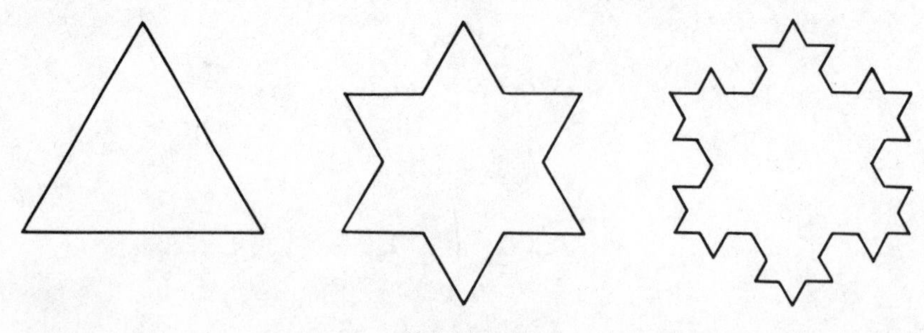

FIGURE 5–5
To make a Koch snowflake, you start with a triangle, then add triangles to each side, and continue to add triangles to the resulting borders.

the area of circle that circumscribes the first step (the six-pointed star) or the area of the hexagon inscribed in that circle, since the interior of both of those regions will also enclose the complete snowflake. Suffice it to say, the area enclosed by the snowflake must be finite, while the perimeter of the snowflake must be infinite.

Mathematics gradually absorbed the pathological curves. By the 1930s, to judge from popularizations of mathematics written then and later, mathematics was proud of its monsters. For the working mathematician, however, the details were generally forgotten. All the mathematician needed to know was that it was important to be careful, because somewhere there existed curves with no tangents or that had infinite lengths in finite regions. Definitions and theorems were made carefully to exclude the monsters. Otherwise, no thought was given to them.

That is, no thought was given to them until 1975 when Benoit Mandelbrot began to publish his thoughts on self-similarity and dimension. The Koch snowflake is the ancestor of a whole class of figures that have self-similarity. Look again at the drawing. Unlike the rectangle, which changes size at different scales, the infinite-sided snowflake looks exactly the same at any magnification that is a multiple of three.

Mandelbrot observed that most natural surfaces and borders are more like the Koch snowflake than they are like the standard cylinders and triangles of the geometer. He pointed out that "clouds are not spheres, mountains are not cones, coastlines are not circles, and bark is not smooth, nor does lightning travel in straight lines." He began to work out a mathematical treatment of natural curves and surfaces, many of which are in a certain way self-similar, for which he coined the term *fractal geometry.*

The classic example of a border in nature that resembles a Koch snowflake with respect to self-similarity is the shoreline of a continent or island. At one level, there are occasional large bays or promontories. Enlarge the scale by a factor of, say, 10 and you are looking at a portion of the original coastline only. Within this portion, there are still bays and promontories, although not the same ones as were visible at the original scale. Continue to enlarge the scale until you are looking at just a few meters of shoreline. Even a few meters will show the type of indentations and exdentations that were also apparent when the whole shoreline was visible. In fact, if given no other clues and no knowledge of the scale being used, you could not tell whether you were looking at the shoreline of a nation, of a state, of a county, or of a town (Figure 5–6).

Of course, a coastline is not self-similar in the same sense as a Koch snowflake. Although you cannot tell which scale is which in looking at a coastline, the figures at different scales are not

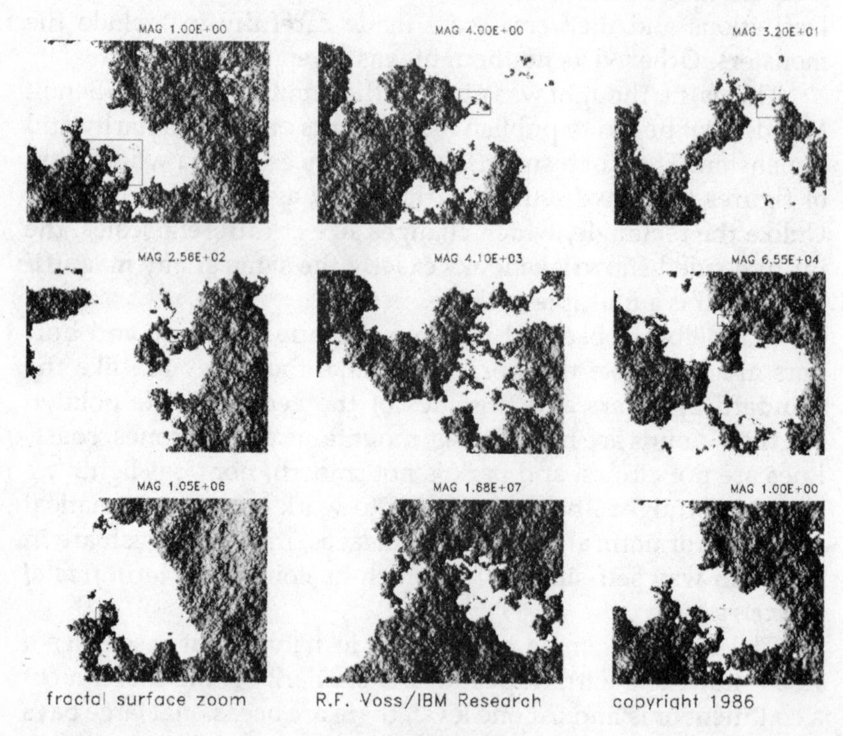

FIGURE 5–6
The shoreline created mathematically by Richard Voss is like a real shoreline in that it looks about the same at any scale.

strictly identical. Natural figures are *statistically self-similar,* while artificial figures, such as the Koch snowflake, are purely self-similar. The property of being statistically self-similar can be rigorously defined, but for our purposes it suffices to say that all statistically self-similar figures have *fractal dimensions* that are not integers (except for the line, of course). Before that can mean much, however, we can use the idea of pure self-similarity to define fractal dimension.

In the earlier discussion of dimension, dimension was tied largely to locating points in space. That approach is pretty much a dead end as far as investigating the dimensions of figures. Instead, you need to have a definition of dimension that is based on properties of the figure being examined. Consider a line segment, a square region, and a cubical region. (A line segment is that part of a line between two points, including the two points; a region is that part of space bounded by a figure, including the figure.) You need a definition of dimension that will give the segment a dimension of 1, the square region a dimension of 2, and the cubical region a dimension of 3; furthermore, this definition must be something that will generalize to other figures. Simply saying that a line segment has length, while a square region has both length and breadth, and so forth, is not sufficient.

Instead, a workable definition can be based on the way that a given figure can be separated to produce self-similar fragments. Notice that a line segment can be separated into any number of segments of the same length and all of these will appear to be identical to the original segment at some appropriate scale. Choose any number r as the ratio of scaling you want to use to show self-similarity. You want to see the same figure when you increase the magnification by r. In that case, the number of self-similar segments you get can also be found as r. For example, if your original line segment is 100 cm long, and you choose r as 10, you will obtain 10 segments similar to the original one. If N is used for the number of segments, the formula for obtaining N from r for a line segment is $N = r$, which can also be written as $N = r^1$.

How does this idea apply to the square region or the cubical region? Consider a square region 100 cm on a side. If you want little square regions that appear the same with a magnification of 10, how many do you get? Make a sketch. It is clear that there must be 100 such square regions. Try another ratio, say 2. You will get 4 square regions that are similar to the original square region. The behavior is defined by the formula $N = r^2$. Similar reasoning shows that for a cubical region, the formula becomes $N = r^3$.

The exponent for r in this formula matches the first criterion for dimension—although it seems to apply to figures with self-similarity only. Moreover, it admits to generalization for figures with self-similarity: If you agree that r is a ratio that will produce N similar figures, then $N = r^D$ would define the dimension D.

Consider the Koch snowflake as a specific example. Focus on one part of the snowflake border. To obtain similar figures, you need to use a magnification, or ratio, of 3. At that magnification, you will obtain a figure that is 1/4 of the original part of the border, so there will be 4 copies. The formula then says that the dimension D is the solution to the equation $3 = 4^D$. To solve such an equation, you can take the logarithm of both sides. You obtain $\log 3 = \log 4^D$, which can be rewritten as $\log 3 = D \log 4$, so $D = \log 4/\log 3$. This fraction can be evaluated with a handheld calculator. It is about 1.26. Thus, the fractal dimension of the Koch snowflake is 1.26, somewhat more than the expected dimension of a curve in a plane, which is conventionally thought of as 1. In fact, Mandelbrot's definition of a *fractal* is (roughly) any figure for which the fractal dimension is greater than the conventional dimension. I say "roughly" because Mandelbrot specifies the type of conventional definition more rigorously than we need to go into.

With this definition of dimension, we can see that for many purposes, dimensions that are not whole numbers have a real meaning for self-similar curves. The same idea can be extended to the statistically self-similar curves that are commonly found in nature. This has enabled mathematicians and physicists to analyze complex phenomena with greater precision. One unexpected extension has been the development of ways to study turbulence, for example—an extremely important phenomenon for any passage through a fluid, such as the passage of an airplane or spacecraft through the atmosphere.

Recent motion pictures have involved the creation of scenes far away from Earth based on fractals. A still photograph that is one of the best-known examples of this computer art is the fractal planet rise created by Richard F. Voss of IBM's Thomas J. Watson Research Center (Figure 5–7). The planet is a sphere, of course; but the continents on the planet were generated and colored randomly using a fractal dimension of 2.5. The planet is rising above a lunarlike landscape that was randomly created with a fractal dimension of 2.2.

FIGURE 5–7
By using different fractal dimensions in a computer program, Voss created the illusion of a planet rising as seen from its moon.

A Very Special Theory

The special theory of relativity, as was noted in Chapter 3, results from symmetry. "The entire theory of relativity . . . is but another aspect of symmetry," as physicist Hermann Weyl once put it. Specifically, the symmetry *idea* is that an observer in one field of reference should not be able to determine whether the given field of reference is moving with respect to a stationary field; or whether the observer is in a stationary field of reference and the observed other field actually is moving. Why is this identified with symmetry, since there are no mirrors involved, even metaphorically? There are two equally valid ways to answer that question. One is to note that if there are two observers, neither one would be able to tell which was moving. The situation is symmetrical in a broader sense of the word than the purely geometrical meaning. In fact, physicists often use such broad notions of symmetry, or even more general ones, as part of their physical reasoning.

The second reason to identify with symmetry the problem of determining which of two moving frames of reference is actually at rest is that the problem can be described with transformations. As we have seen, transformations are intimately connected with symmetry concerns.

You have no doubt experienced this problem yourself at one time or another. If you are a passenger caught in a traffic jam next to a large truck, smooth motion by your car in the forward direction cannot be separated by observation from smooth motion in reverse by the truck (or the other way around—if your car begins to back up, it will appear that the truck is moving forward). The conditions for this effect are that the motion must not involve bumping up and down on the road. Neither should there be any acceleration that you can feel. Finally, you should not be able to see anything known to be truly stationary—such as a lamppost beyond the truck.

Turning the traffic jam into a more general situation, one can say that two observers moving with a steady velocity with respect to one another cannot determine if indeed either one is actually at rest unless there is some third object observable that is known to be at rest.

Now consider the situation in which you are on one planet and another observer is on another planet. Again, in the absence of any acceleration, you cannot be sure which of you is moving and which is standing still. Of course, we normally think of all planets as moving, which they are with respect to the Sun. But we also think of the Sun as moving though space with respect to the galaxy. It would also appear that the galaxy is moving with respect to other galaxies, all of which also appear to be in motion. But just as you cannot be sure whether your car stuck in traffic is moving backwards or the truck is moving forwards, nothing about the complex motions of the planets, stars, and galaxies proves that you are really moving. Although the mathematics is simpler if you think of the Sun as stationary and the planets as moving, no amount of observation could show that it was not the case that the Earth is stationary and the Sun and all the other planets in motion with respect to it.

In the nineteenth century, scientists began a sophisticated search for stability in this complex universe. Their goal was to locate a single motionless "object" with which all the motions of the universe could be compared. Since heat, light, electricity, and magnetic forces all traveled through a vacuum, various scientists postulated that there must be some medium (or several different media) through which these traveled. They called these media *aethers.* The theory of electromagnetism, worked out so brilliantly by James Clerk Maxwell in 1873, established that only a single aether, called the ether today, would be needed to explain light, heat, electricity, and magnetism. Maxwell's equations described the behavior of electromagnetism in terms of waves. Furthermore, production and detection of radio waves—not to mention the familiar properties of light—established that such waves actually existed. From the general theory of wave motion, it was possible to deduce the kind of medium in which such waves would travel.

Since the ether must exist and since electromagnetic waves travel infinitely unless something absorbs them, all the objects in the universe are either at rest with respect to the ether or are moving through it. Scientists had learned from Copernicus and Galileo the dangers of placing the Earth at the center of the universe, so they assumed that Earth must travel through the ether. If so, measuring the change in the velocity of light in two directions (one somewhat against the ether and the other traveling somewhat with the ether) would tell the actual motion of the Earth. A number of experiments along these lines were set up, of which the most famous (because it was the most careful) was the Michelson-Morley experiment of 1887. Other scientists, notably Armand Fizeau in 1851, used other properties of the effect of ether on light to determine how its velocity would be affected. Fizeau's results were taken at the time to show that ether existed, and a later analysis (by Hendrick Lorentz in 1895) also assumed that ether was needed to explain Fizeau's result. Unlike Fizeau's experiment, the Michelson-Morley experiment failed to detect the ether. Many scientists interpreted this to mean that the ether must be somehow carried along with the Earth.

George Fitzgerald suggested another explanation for the failure of the Michelson-Morley experiment. He thought that distance might contract in the direction of motion. In 1895 he worked out just the transformation that would be needed to account for the velocity of light not changing as the Earth moved through the motionless ether. Lorentz—independently and with something else in mind—came up with the same transformation. As a result, this transformation is generally known today as the Fitzgerald-Lorentz contraction.

Along came Einstein in 1905. He was familiar with Fizeau's experiment and Lorentz's discussion of it. Although whether he knew about the Michelson-Morley experiment or not has been somewhat controversial, it seems to be well-established that he knew about it, but that it was not a prime motivation in his work. He was also familiar with the Fitzgerald-Lorentz contraction, which Fitzgerald and Lorentz assumed was something that happened to matter as a result of absolute motion. Einstein's work in 1905 was to show that the experimental results were all valid, but that their interpretation was clumsy and inappropriate. In fact, the correct interpretation that Einstein supplied explained not only these results but much more.

Specifically, Einstein observed a flaw in the common way of viewing time. If two events occur at different places in the universe, earlier investigators assumed that it was possible to compare both events with some standard clock, thus determining whether or not the events happened at the same time or at different times. Einstein noted, however, that the only way this could be performed was by having a signal from one event reach the other. Normally, such a signal would be that an observer at one event saw the other event occur; that is, the signal would be light from the distant event reaching the observer. This simple consideration results in far-reaching consequences.

In physics before the special theory of relativity, every event could be located with respect to a three-dimensional space (coordinate system) and a universal clock. Thus, each event was in a four-dimensional space whose coordinates could be labeled (x, y, z, t). A different choice of coordinates that used the same units should not change the relationship between two events. For example, if event A was five meters from event B and event A occurred 12 seconds before event B, the distance in time and space between A and B would not change with a change in coordinate systems. To be more specific, let us suppose that event A is dropping a flag and that event B is firing a gun. Furthermore, suppose that both events take place on the deck of a large ship that is smoothly moving through calm waters (Figure 5–8).

The coordinates of A with respect to the deck of the ship can be taken as (1, 2, 0, 0), where we ignore the height above the deck and start the clock when the flag is dropped. Similarly, the coordinates of B with respect to the deck are (5, 5, 0, 12). Now the distance between A and B (taking time into account) is found using the distance formula for a Cartesian coordinate system, which is the positive square root of the sums of the squares of the differences between coordinates. You begin by calculating the square of the distance, D^2. In this case,

$$\begin{aligned} D^2 &= (5-1)^2 + (5-2)^2 + (0-0)^2 + (12-0)^2 \\ &= 4^2 + 3^2 + 0^2 + 12^2 = 16 + 9 + 0 + 144 = 169 \end{aligned}$$

The square root of 169 is 13, so event A is 13 meter-seconds from event B.

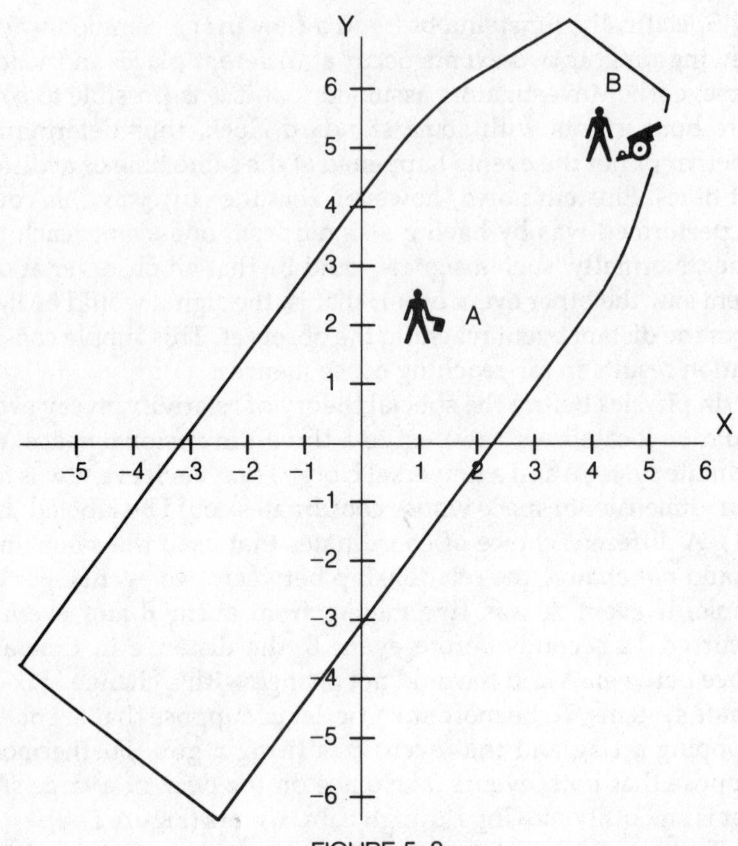

FIGURE 5–8
Twelve seconds after the flag is dropped on the deck of the ship, the cannon that is 5 meters away is fired.

But the ship is moving smoothly through the water. We could also have measured the distance between A and B by using a coordinate system that is embedded in the Earth. Furthermore, we could have started the clock when the ship left port. For the laws of physics to make any sense, measuring the distance using this second coordinate system must also give a distance between A and B of 13 meter-seconds. At any given time, the relation of one coordinate system (deck of ship, clock starts with drop of flag) to the other coordinate system (surface of Earth, clock starts with ship leaving port) can be found by a simple transformation that moves one coordinate system into the other. Such a transformation might require a combination of a translation and a rotation, but we can simplify our computations without a loss of generality by assuming that the two coordinate systems are oriented so that only translation is required. If the translation is moving a units in the x direction, b in the y, c in the z, and d in the t, then the new coordinates of A will be $(1 + a, 2 + b, 0 + c, 0 + d)$, while the new coordinates of B will be $(5 + a, 5 + b, 0 + c, 12 + d)$. Looking back at how the distance was calculated shows that a, b, c, and d will drop out, so the requirement that the distance stays at 13 meter-seconds is met.

Remember, however, that this was what physicists believed (usually without realizing that they believed it) *before* 1905. Einstein pointed out that there was no way to be sure that the times of A and B were the same unless a signal passed from A to B and that such a signal must also take the same time from one coordinate system to another. When light is taken as the signal, which of necessity it must be for physics to make sense, the speed of light must be the same for both coordinate systems. In other words, the Michelson-Morley experiment necessarily found that the speed of light did not change when measured with and against the motion of the Earth. Einstein did not need the experiment to determine that the speed of light was constant. He had worked it out from general principles, specifically from the symmetry consideration that there should be no change from one coordinate system to another moving smoothly (technically, without acceleration) with respect to the first.

An additional result of taking the signal into account is that the idea of an independent clock must be abandoned. Instead of having a clock that is independent of the events, which assumes absolute time, each observer is given a clock. Einstein's time system abandons the idea that you can tell what "at the same time" means in general. Instead, you can only tell what "at the same time" means for a particular point in space. At that point, you can measure when the signal from another event arrives and compare it with the time on your clock. Thus, the space-time distance between two events must take the speed of light into account. The way to do this is to multiply each time by the speed of light. Including this factor in the distance formula gives the result that the difference in the two times is multiplied by the square of the speed of light. In other words, if we use c to mean the speed of light, the square of the distance between A and B is $16 + 9 + 0 + 144c^2$. This is the quantity that must stay the same whether it is measured from the deck or from the surface of the Earth.

When the speed of light is taken into account, a simple translation no longer suffices to keep the space-time distance the same between two moving coordinate systems. Instead—and this really should not be a surprise—the transformation required turns out to be exactly the Fitzgerald-Lorentz contraction. Again, Einstein determined an important result that had been developed only because it fit experimental results without even taking the experiment into account. The result had to be believed because it came from the simple consideration of what it meant for two events to happen at the same time, provided that essentially symmetrical results be obtained. Symmetry in this context means that if you change from one coordinate system to the other, you should not get different results. That is, the thing on one side should look just like the thing on the other.

Later, Einstein considered the case where one coordinate system was accelerated with respect to the other. The symmetries involved in that consideration led to some rather difficult mathematics and to the general theory of gravitation, known more often as the general theory of relativity.

The concept of symmetry that is used in the theory of relativity is one of the commonest in physics. It is both the simplest and the deepest, as will be evident many times in the remainder of this account of symmetry. People who are familiar only with line symmetry often do not recognize that any situation in which you can exchange two entities without changing anything else about a situation is a form of symmetry. Indeed, the physicist Edward Teller has noted "The word 'relativity' is a bit unfortunate because in Einstein's theory, the main point is not what is relative, as most people believe, but rather what is invariant. . . ." in the case of the universe, the speed of light. Invariance is the essence of symmetry.

6

Predicting the Unknown

Symmetry is a basic principle of physics. There are deep philosophical, mathematical, physical and psychological reasons for this, and physicists will persist in looking for some phenomenon predicted by a theory based on symmetry even when it begins to seem a little absurd to continue.

Dietrick E. Thomsen

"Of course, they answer to their names?" the Gnat remarked carelessly.

"I never knew them to do it."

"What's the use of their having names," the Gnat said, "if they won't answer to them?"

"No use to them," said Alice; "but it's useful to the people that name them, I suppose. If not, why do things have names at all?"

"I can't say," the Gnat replied. "Further on, in the wood down there, they've got no names—"

Lewis Carroll

One of the remarkable things about the special theory of relativity is that Einstein was able to obtain results such as the constancy of the speed of light (in a given medium) and the Fitzgerald-Lorentz contraction without doing experiments. In fact, the Fizeau experiment of 1851 was, at that time and even in 1905, interpreted as disproving the constancy of the speed of light; but Einstein ignored this. Instead, Einstein discovered the special theory and established these results (and others) by using symmetry arguments about the physical situation followed by a mathematical development.

There have long been two mysteries about the relation between mathematics and physics. One of them concerns the ability of mathematicians to invent the mathematics needed for a physical theory from completely abstract principles well before anyone has thought of the physics involved. An example of this is that the mathematician Bernhard Riemann created in 1854 the abstract geometry that Einstein needed for the general theory of relativity in 1916. None of the explanations offered for such mathematical anticipation have ever been wholly satisfactory. It is such a general phenomenon, however, that many of today's mathematicians serenely work on abstract problems that seem to have no known use. They are serene because experience so far shows that no matter how peculiar it seems, good work in mathematics will eventually be put to use in science, generally in physics.

Morris Kline has a slightly different viewpoint. Professor Kline suggests that Riemann's geometry was not necessarily the best mathematical expression for the general theory. Instead, Einstein looked around for some kind of mathematics that could handle his new physical ideas. Riemann's work (as developed by several subsequent mathematicians) seemed to be the best thing available, although far from perfect. Similarly, Kline points out that Kepler's use of the geometry of the ellipse to describe the motions of planets was also a shoehorn job. Given the observations, the ellipse was the nearest curve that would fit. But in real life, planets have moons and are also affected by the gravitational field of other planets. As a result, none of the planets actually travels in an ellipse. Kline's point is that the mathematics developed in one age is not magically the description needed for the physics of the next age; instead, the physicist looks through all of mathematics (or at least he or she could do that) and chooses the mathematical description that works.

The other mystery is the one with which we are concerned with here. It is that mathematics, especially mathematics that is based on symmetry concerns, can extract from a few known physical facts glimpses of reality that are totally unexpected. Einstein's special theory of relativity did not rest on experimental observation. In the case of the special theory, however, the experimental observations existed. Sometimes, however, the theory can be far in advance of experiment.

Anticipating Reality

One of the most dramatic of such results is the prediction of a phenomenon that has never before been observed. This has happened repeatedly in modern physics. Examples from astrophysics, for example, include black holes, pulsars, gravitational lenses, and the leftover radiation from the Big Bang, all of which were predicted before they were observed.

One particular manifestation of this kind of prediction was the 1930 anticipation of antimatter by Paul A. M. Dirac, two years before the first particle of antimatter was observed by Carl D. Anderson. Both physicists won separate Nobel Prizes for these achievements, but I have always felt that the more remarkable story is Dirac's anticipation, based—as you might expect—on symmetry principles extended by appropriate mathematics. This is an "anticipation" rather than as a "prediction" because Dirac did not recognize the possibility that his mathematical results might turn out to indicate a previously unknown form of matter. Instead, he tried to make ordinary matter fit the solutions to his equations, an effort that failed. In this case, it took an experimental discovery to put the correct interpretation on what Dirac had found.

We begin with another general way to look at symmetry in algebra. Recall that a function can be even or odd. Parity for functions is a special case of this more general notion of symmetry in mathematics. If you can interchange the variables in an equation in a systematic way without altering the equation (in some fundamental sense, which may vary with the situation), the equation is symmetric. An easy example to contemplate would be an equation such as $x + y = 1$, which is the same equation for all practical purposes as $y + x = 1$. A more complex example is an equation from trigonometry called the law of cosines, a rule that describes the relation among three sides of a triangle and the angle between two of those sides (Figure 6–1).

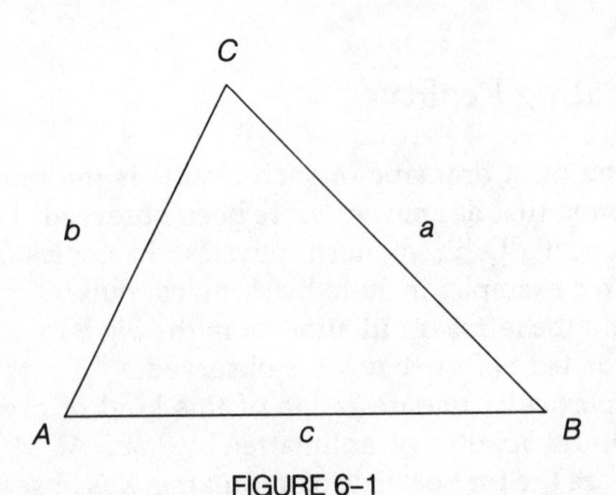

FIGURE 6–1
By convention a triangle with vertices *A*, *B*, and *C* has sides of lengths *a* opposite *A*, *b* opposite *B*, and *c* opposite *C*.

Conventionally, the lengths of a triangle's sides are labeled with lower-case letters that match the capital letters of the angles opposite them. Thus, the side opposite angle C is labeled *c* and so forth, as shown in the diagram. With this convention, the law of cosines can be stated as

$$c^2 = a^2 + b^2 - 2ab \cos C$$

(It should be noted that the law of cosines is just a generalization of the Pythagorean Theorem, since if the angle C is a right angle, $\cos C = 0$.) Since the triangle used was arbitrarily labeled, it is not surprising that the algebraic equation for the law of cosines is symmetric. If you interchange *a, b,* and *c* in a systematic way, you obtain two other versions of the law of cosines that are equally valid. The way to do this is to move each variable one place to the right, with the one on the right-hand end moving as if in a circle back to become the first variable. In other words, *c* replaces *a, a* replaces *b,* and *b* replaces *c.* At the same time, angle C must be replaced with its corresponding angle, which will be angle B, since *b* replaces *c.* The new form of the law of cosines is

$$b^2 = c^2 + a^2 - 2ca \cos B$$

That this form (and the other one that can be produced by one additional interchange) is also true is evident by thinking about the initially unlabeled triangle, which could have been labeled this way at the start. Note that a similar interchange for the Pythagorean Theorem using conventional labeling would produce $b^2 = c^2 + a^2$, not a valid form of the theorem. The reason is that *c* has a privileged position in the conventionally labeled right triangle being described by the Pythagorean Theorem; it is always the length of the *hypotenuse,* the side opposite the right angle. So the symmetry of the equation for the law of cosines comes from the symmetry of the initial situation in which none of the sides nor angles of the triangle was privileged.

Related to this is the kind of symmetry we looked at with even and odd functions. For an even function, changing the variable to its negative counterpart does not change the function. The same can be true of an equation that does not represent a function. For example, the law of cosines has three kinds of symmetry based on changing variables to their negatives. In the original form, the variable c is not symmetric to the other two variables; that is, it occupies a special place with respect to them. This is evident if you replace c with $-c$ and replace angle C with its negative. Since $(-c)^2$ is the same as c^2 and since the cosine is an even function, meaning that cos $-$C is the same as cos C, the equation does not change in any way. However, if you replace either a with $-a$ by itself or b with $-b$ by itself, you will get a new equation that is equivalent to

$$c^2 = a^2 + b^2 + 2ab \cos C$$

which is not the same equation at all. On the other hand, you can replace both a and b by their negatives (the second form of symmetry) or all three variables and the angle with their negatives (the third form of symmetry). In either case, you will get an equation that is equivalent to the original one.

Dirac was a physicist who had trained as a mathematician. In fact, all of his formal education was in electrical engineering or in mathematics, although his subsequent career was entirely in physics. Either because of this education or because of a natural predilection, Dirac's work in physics is mathematically quite elegant.

The situation that faced particle physicists in 1928 was that although the particle theory known as quantum mechanics was successful, this theory took no account of Einstein's special theory of relativity. Such a situation might have been acceptable if electrons and protons (the neutron was not to be discovered until 1932, the same year as the discovery of antimatter) just sat around. In fact, however, such particles were often to be found whizzing around at speeds close to that of light in a vacuum, the ultimate speed of special relativity. The electron, in particular, behaved as if it were a particle moving rapidly in an orbit when it was in an atom; and in other manifestations, electrons were also found speeding through cathode-ray tubes or generally doing anything but just sitting around. Even the electrons of "static" electricity are actually moving.

Dirac set out to develop a theory of the electron that would resolve the differences between relativity theory and quantum mechanics. Others had tackled this difficult mathematical task and failed, including such luminaries as Erwin Schrödinger. Dirac, however, succeeded. Dirac discovered a new version of the Schrödinger wave equation that included a term to account for the relativistic effects, and, behold, spin emerged. So spin was not a mysterious quantity at all, but instead was a direct consequence of relativity. This was very satisfying, although quite unexpected.

The mathematics that resulted from Dirac's new formulation was formidable, and, furthermore, the mathematical consequences that emerged were daunting. For one thing, calculations kept resulting in infinite amounts for quantities known to be finite, such as the electron's mass and charge. It took 20 years and three other brilliant mathematical physicists to find a way around this dilemma. No one was willing to give up the theory, because it was right much of the time and logically necessary; but no one wanted those infinite quantities as answers, either.

The other problem was that the new version of the wave equation was symmetric in a way that the original Schrödinger version was not. In fact, Dirac remarked that if Schrödinger had realized that his original equation was not sufficiently beautiful, then he would have discovered the correct, beautiful (because symmetric) equation. The new symmetry required that both a positive and a negative solution for the energy of the electron must exist.

Here we are talking about *energy* in the sense of Einstein and relativity; that is, the energy of the electron is a combination of its mass and energy as defined in prerelativistic terms. This is *not* to be confused with negative *charge.* Physicists were quite used to the negative charge of the electron and the positive charge of the proton. Negative energy was something else. Negative energy did not make sense to anyone.

Dirac tried to get rid of this unwelcome symmetry. But the consequences of discarding the solution with negative energy were so unpleasant that this option was ruled out.

After a couple of years of worrying about the problem, Dirac published his proposed solution to it in 1930. Dirac noted that if these states of negative energy existed, electrons would "roll downhill" into them. Because this was not observed, something must be keeping electrons from falling into the states. That something must be the Pauli Exclusion Principle. In other words, if almost all of the negative energy states already contained electrons, then the electrons we observe would be unable to "roll downhill." If you think of the observed electrons as rocks on a slope, Dirac implied that the valley below was already filled with rocks. There was no place to go.

This idea has many implications, not all of which were understood in 1930. Essentially, Dirac was proposing a new concept of the vacuum. The quantum vacuum is not the same as the classical vacuum. In the classical vacuum, nothing ever happens because nothing is there. In the quantum vacuum, there is always a finite probability that something will happen, because a particle can rise out of this sea of negative energy like a catfish jumping in a pond in summer. We don't get much of a chance to observe either the creation of a particle or the jumping catfish because both return to their sea or pond very quickly. Nevertheless, both events happen and they, or their effects, can be observed. For example, you hear the catfish land in the water and can see the ripples, even though you usually do not see the fish. Dirac was right about the quantum vacuum, although this was not immediately clear even to Dirac.

Dirac also found that a few states would be unfilled by the sea of electrons. What would they be like? This problem had already been addressed by particle physicists in a different situation. The earlier problem was what would happen if you removed an electron from one of the inner orbits of an atom. It had been worked out that the result would be the same as if there were a positive charge occupying the same orbit as the removed electron. So Dirac predicted that the "hole" that existed when a state of negative energy was unoccupied would be perceived as a particle with positive charge.

Of course, you should not be surprised that a hole can take on a life of its own. We observed that in Chapter 5 when we learned ways to deal with the 15 puzzle. Following the path of the hole was more useful than following the path of any of the tiles.

At this point, Dirac uncharacteristically made a mistake. Since the only subatomic particles known in 1930 were the electron, proton, and photon, Dirac tried to force one of these particles into being the hole. This would have been quite satisfying, since everyone agrees that the fewer the particles the better. The only possible candidate was the proton, because it has the right charge. The proton, however, is about 2000 times as massive as the electron, which does not seem to be very symmetrical. This worried Dirac and he ingeniously argued that such a dissymmetry should be expected—but he was wrong. Almost immediately, Robert Oppenheimer pointed out the mistakes in Dirac's argument. But Oppenheimer did not think that the "sea of negative energy" concept of the vacuum should be dismissed.

So the matter rested for two years. In 1932, Carl D. Anderson was trying to find particles that reached the ground, but which had their origins in outer space. Such particles were called *cosmic rays,* and there are really two quite different types. The first kind, called *primary* cosmic rays, are the actual particles that reach the Earth from space. Mostly these are fast-moving protons or nuclei of atoms. *Secondary* cosmic rays are formed when the primaries collide with atoms in the atmosphere. Most of the cosmic rays that reach ground level are secondaries. Carl D. Anderson and Robert A. Millikan, both at Cal Tech in 1932, designed a device to trap and identify the secondary cosmic rays. It consisted of a chamber in which a charged particle produced a visible track by forming ions in a cloud of alcohol—a cloud chamber. The chamber was encased in a thin layer of lead to keep out charged particles with less energy than that expected for cosmic rays. The cloud chamber was also subjected to a strong magnetic field, which would cause the paths of the charged particles to bend. Finally, there was a camera, so that the faint curves of the vapor trails could be recorded.

On August 2, 1932, Anderson's camera recorded something unexpected. A particle came zipping into the cloud chamber, passed through a thin lead plate, and continued on for a total path of about 5 centimeters. The direction of the particle through the chamber could be determined because it had more energy on one side of the lead plate than on the other. The amount of energy could also be determined by the amount of curvature caused by the magnetic field. With the direction of the particle known, the direction of the curvature could be used to reveal the charge of the particle. It was positive. This presented Anderson with an apparent paradox. The only positive particle known at the time was the proton, and the path of a proton in that particular situation would be about 5 millimeters instead of 5 centimeters. So the particle was not a proton, and, since the charge was positive, it was not an electron. By that time, Chadwick had discovered the neutron, but the essence of being a neutron was to have no charge, so that was out.

Measurement showed that the range of mass and charge must be within a reasonable proximity to that of the electron (although with opposite charge). Anderson commented, "It is concluded, therefore, that the magnitude of the charge of the positive electron which we shall henceforth contract to positron is very probably equal to that of a free negative electron which from symmetry considerations would naturally then be called a negatron." (Symmetry does not always win out; while the name *positron* was kept, no one calls an electron a *negatron.*)

It was not recognized immediately that Anderson's positron was the particle predicted by Dirac. Eventually the connection was made. Dirac later remarked that he was surprised that his equation had turned out to be so much more intelligent than he was himself.

If enough energy was applied to the apparent vacuum, which Dirac had called a sea of electrons, one of the electrons could gain enough energy to rise from the sea. This electron would leave a hole behind. What we perceive at the hole's location, however, is a region of positive charge that has the same mass as an electron, or a positron. Thus, it would appear that an electron and a positron were created out of nothing.

Actually, this can happen in two different ways. The energy could be supplied by a photon. When that happens, the electron and the positron can travel quite far through space. The electron, in most circumstances, will travel on indefinitely unless it bounces into something (such as another electron) or is captured by a region of positive charge (such as an ion). The positron, in most circumstances, will not travel very far. Soon the positron will encounter an electron. From the point of view of Dirac's 1930s theory, the encountered electron will "fall into the hole." What we perceive is that the electron and positron annihilate each other, producing an energetic photon.

These two events—the creation of the electron-positron pair and their annihilation—can be diagrammed using a simple method invented by Richard Feynman and used today by all particle physicists (Figure 6–2).

The Feynman diagrams show that the two events are essentially the same. You can think of the reaction as purely time-symmetric, which is true of most reactions in physics. That is, if you reverse time for the first event, you obtain the second event;

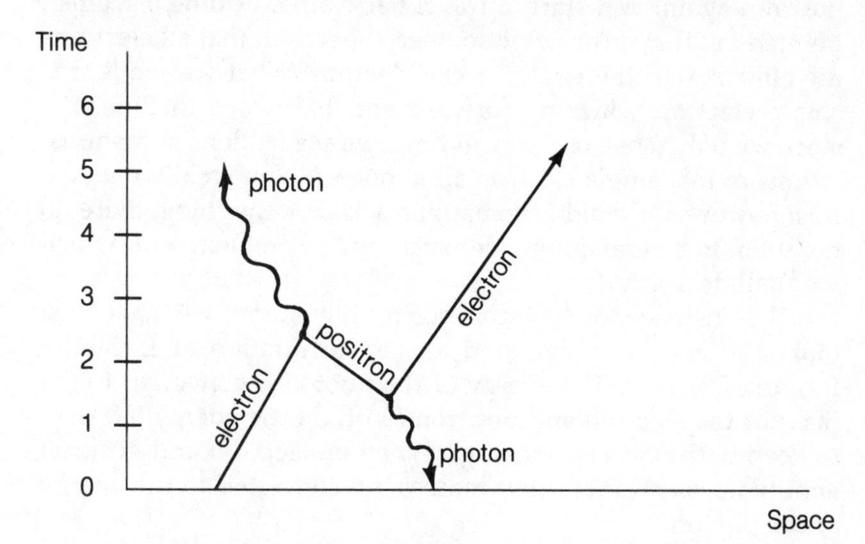

FIGURE 6–2
A Feynmann diagram showing an electron turning into a positron after absorbing a photon looks just like a diagram showing a positron meeting an electron and disappearing into a photon.

by time symmetry, both events should be possible, and, in fact, they both occur.

Feynman has made a light-hearted, but somehow compelling suggestion that time reversal could occur in another way, a way that would explain one of the mysteries of particle physics. The mystery is why the electron always has the same mass and the same charge (and why the positron also has the same mass and charge as the electron). There seems to be no known physical law that would require all electrons to be identical in this way. Feynman noted that his diagrams could also be interpreted to mean that a positron was simply an electron traveling backwards in time. In that case, the diagram would show an electron approaching a point in space while moving forwards in time, absorbing a photon, and reversing time to travel backwards in time as a positron. After a short distance, the positron would emit a photon, reverse time, and travel forwards in time a long period as an electron. Eventually, it would absorb a photon, become a positron again, and start to travel backwards in time. Feynman pointed out that if this were so, then the reason that all electrons are alike is that there is only *one* electron. What we see is that single electron whizzing forward and backward in time, but since we only move one way in time, we see trillions of manifestations of that single electron all at once. No one really believes this, however; nor did Feynman himself. For one thing, there do not seem to be enough positrons around, a problem with which we shall deal shortly.

The instance of electron-positron creation we have just looked at can be understood as a manifestation of Einstein's formula $E = mc^2$. The energy of the photon is converted to the mass of the electron and positron (with a little energy left over to keep both particles moving). When an electron and positron annihilate each other, their masses are converted to the energy of the photon.

A different instance of electron-positron formation seems more mysterious. In that case, the electron-positron pair is formed from random fluctuations of the vacuum. A modern definition of a vacuum is the state of least energy for a region. We now know that this energy can be the lowest *average* energy. Even when the average energy is 0, which we would think of as a normal vacuum, there can be fluctuations above and below the average. For example, an electron-positron pair can appear briefly and then annihilate each other. This can be viewed as a random fluctuation of energy that is permitted to break the laws of conservation of energy because it is so brief. In fact, it is like successfully kiting a check (even if it is illegal). You know that on Tuesday you can write a check at the grocery store even when you have no money in your bank account provided you deposit money in the account on Wednesday. Similarly, the electron-positron pair can violate the law of conservation of mass-energy as long as it makes it up soon enough that nobody finds out. How soon is that? The Heisenberg Uncertainty Principle gives specific limits to our ability to observe something. As long as the electron-positron pair disappears before those limits are reached, everything is hunky-dory.

You may be a bit skeptical about this. After all, if such a creation of matter out of nothingness cannot be observed, even in principle, why should anyone believe that it happens. Oddly enough, this peculiarity of the vacuum produces measurable effects, specifically a change in a quantity known as the Lamb shift. Since these effects *can* be observed, we accept the idea that electron-positron pairs form and rapidly disappear in the best vacuum money can buy. Again, it is like the unobserved catfish and the observed splash.

The Lamb shift is a bit difficult to explain. A look at a similar phenomenon of the vacuum will show how something that cannot be directly observed can produce an effect that is directly observed. Not only do electrons and positrons arise and disappear as a result of random fluctuations in the vacuum, but photons also appear and disappear. Photons are even particles, unlike electrons and positrons, which are odd, so photons do not behave at all like electrons. Most of the time it pays to think of odd particles as particles and even particles as waves, although each are both. So think of the creation of transitory photons in the vacuum as the production of electromagnetic waves out of nothing. These waves cannot usually have a very high energy, however, since they are being created from nothing and will quickly disappear into nothing. Waves with a low energy have a long wavelength, while waves with a high energy have a short wavelength; hence, most waves that create themselves from nothing will have long wavelengths.

Consider the following situation. Two plates of metal are placed close together in a vacuum. On the outside of the two plates, the vacuum is producing electromagnetic waves (with such a short lifetime that they are not observable) of various long wavelengths. Some wavelengths may be fairly short, but most will be long. Between the two pieces of metal, there is no room for a long wavelength. There is only room for short wavelengths. Not very many of these will be created and the few that are will sink back into the vacuum very quickly, since the more energy it takes to create the wave, the shorter a time it can last before it would be observed, which would not be allowed. This rule is like a rule for kiting a check such that the greater the sum the check is written for, the sooner it will reach the bank. Fortunately for the economy of most individuals—and some corporations—this is not the case in the banking world.

The result of the difference between the part of the vacuum outside the two plates and the part inside the plates is this: There are many more electromagnetic waves that transitorially exist outside the plates than between the plates. These exterior waves create a pressure that pushes the plates together. This pressure can be measured. Thus, even though we cannot observe the actual electromagnetic waves created in the vacuum, we know they are really there.

This view of the vacuum gradually replaced Dirac's concept of a sea of electrons. The modern concept of an active vacuum does not require a sea of electrons to explain the solutions to the Dirac equation. Physicists are pleased with this because, as Victor Weisskopf has remarked, the idea of a sea of electrons is ugly. This idea is "ugly" because it assumes something that is, by its nature, totally undetectable is found everywhere in the universe. It complicates things, rather than making them simpler.

Is There an Antigalaxy?

When Dirac's anticipation of the positive electron, or positron, was confirmed, physicists realized that his interpretation of the solution of the relativistic wave equation must be essentially correct. In that case, the proton, which obeys the same kind of relativistic wave equation must also exist in another form. There must be a negatively charged proton. This particle, like the familiar proton in every way except for charge, came to be called the antiproton. By what linguists call a "back formation," the positron acquired its third name, *antielectron.*

Furthermore, if one could obtain an antiproton and set an antielectron orbiting about it, the result would be antihydrogen, the simplest element of antimatter. The first hitch in this scenario, however, is obtaining the antiproton. Calculations show that while only a million electron volts (1 MeV) of energy is needed to pop up an antielectron from the vacuum (or, in Dirac's earlier, "ugly" picture, to raise a negative-energy electron to a state of positive energy, leaving a hole that behaved as the positron), the absolute lower limit of energy needed for the antiproton is nearly two billion electron volts. The problem is the proton's mass. Both the proton and its twin, the antiproton, have a mass that is about 2000 times that of the electron. A direct result of Einstein's famous formula $E = mc^2$, which relates mass (m) and energy (E), is the energy requirement. Since a proton and an antiproton are to be produced from nothing but energy, the energy required is twice the product of the mass of the proton times the speed of light (c) squared, or about 1.88 billion electron volts (1.88 BeV).

In practice, about six billion electron volts (6 BeV) were thought to be needed to guarantee production of a detectable antiproton. It seemed unlikely that an antiproton would be found as a result of a chance event with an unusually powerful cosmic ray. Instead, the antiproton would have to be created in a particle accelerator (known commonly, but misleadingly, as an "atom smasher"). In the 1950s, a large particle accelerator was planned and built with the 6 BeV requirement in mind. In 1955, the particle accelerator, called the Bevatron, was able to produce the required energy. Emilo Segrè and Owen Chamberlain used the Bevatron to produce about 250 antiprotons, for which they won the Nobel Prize in Physics.

To make antielements beyond hydrogen would require an antineutron as well as an antiproton. Atomic nuclei more complicated than that of hydrogen (whose nucleus is just a proton) form only as mixtures of protons and neutrons. Protons in combinations with other protons—and, by analogy, antiprotons with antiprotons—are too unstable to form nuclei. But there would seem to be a problem in obtaining antineutrons. The difference between an electron and an antielectron or a proton and an antiproton is that the antiparticle has the opposite charge of the particle. A neutron has no charge. Yet there are other quantities that can be "either/or" that have parity. One of these is spin, which can be either clockwise or counterclockwise as viewed from above the "north pole" of the neutron.

What is the north pole of the neutron? The neutron has a magnetic field, albeit a small one, that is similar to that of the Earth. A closer look at the theory of antiparticles reveals that charge is *not* the only difference between a particle and its antiparticle after all. In addition, a reversal of spin (as seen from a specific pole, either north or south) occurs. If the particle spins clockwise as seen from above a given pole, the antiparticle from the same position will be seen to spin counterclockwise. An equivalent way to think of this—one that is actually more common among physicists—is to keep the spin in the same direction and reverse the magnetic field. Thus, even though it has no charge, you can tell a neutron from an antineutron.

The preceding paragraph, while providing an easy way to picture spin, is somewhat removed from reality, since it suggests that you could actually see a subatomic particle spinning. Suffice it to say that this mental picture is an oversimplification of a mathematical reality.

Segrè and Chamberlain used the power of the Bevatron to produce antineutrons. Somewhat later, the antiproton and antineutron were joined by Leon Lederman and coworkers to form the nucleus of an atom of heavy hydrogen. Such a particle, called the *antideutron,* established that antimatter beyond ordinary hydrogen could actually exist.

As particle accelerators gained far more energy than the Bevatron, it became clear that antiparticles existed for every particle. In fact, according to the rules developed for classifying particles and antiparticles, the familiar subatomic particle called the *neutrino* is not a neutrino at all—it is an antineutrino. Since an antineutrino is produced when a neutron decays, a neutrino would be produced by the decay of an antineutron. Other particle decays produce neutrinos, such as one decay mode of a pion or positron emission by a nucleus, but the original neutrino, now known to be an antineutrino, was defined in terms of neutron decay.

When a particle seemed not to have an antiparticle, such as the photon, physicists decreed that it was its own antiparticle. This is in analogy to the role of 0 among the signed numbers. Every number has its opposite or antinumber. The antinumber for +3 is −3. The antinumber for −5 is +5. But the antinumber for +0 or −0 is just 0, since all three symbols refer to the same number. Similarly, the photon is its own antiparticle in the same way as zero is its own antinumber.

Furthermore, for most sufficiently energetic reactions produced in particle accelerators, particles and antiparticles are produced in equal amounts. It is not necessary to set up special experiments to get antiparticles. They are a result of the symmetry that appears at higher energies (of which more will be said later).

This presents a major puzzle. If symmetry considerations suggest that antimatter and matter should be present in the same amounts at high energies, where is the antimatter? We now are quite certain that the universe was created in a very high energy explosion known as the "Big Bang." Such an explosion should have produced equal numbers of particles and antiparticles. Yet, all around us electrons, protons, and neutrons are common, while antielectrons, antiprotons, and antineutrons are exceptionally rare. What became of the antimatter?

As early as 1933, Dirac was already puzzled by this problem. In his Nobel Prize address he noted, "If we accept the view of complete symmetry between positive and negative electric charge so far as concerns the fundamental laws of nature, we must regard it rather as an accident that the earth (and presumably the whole solar system) contains a preponderance of negative electrons and positive protons. It is quite possible that for some of the stars it is the other way about, these stars being built up mainly of positrons and negative protons. In fact, there may be half the stars of each kind. The two kinds of stars would both show exactly the same spectra, and there would be no way of distinguishing them by the present astronomical methods." Dirac certainly believed in pure symmetry.

It is clear that there would have to be some separation between pockets of matter, like the solar system, and the proposed pockets of antimatter. Antimatter cannot exist in the presence of matter. As in Dirac's "sea of negative energy" concept, the meeting of an electron and an antielectron is the equivalent of the meeting of a hole and an object of exactly the same size that falls into the hole. The hole disappears, because it is filled with the object. The object disappears, because it is in the hole. The principal difference between actuality for electrons and antielectrons and this hole analogy is that an electron falling into an antielectron hole releases energy in the form of photons. Since the energy of the electron and the antielectron cannot simply disappear, the photons take care of it. Similarly, when other antiparticles meet their corresponding particles, they also annihilate each other and produce energy. Although the immediate result of a proton meeting an antiproton is a burst of particles that have mass, these particles very quickly decay into either neutrinos or equal numbers of electrons and antielectrons. The electrons and the antielectrons then annihilate each other, producing photons. Thus, the ultimate result of any encounter between a particle and its antiparticle is pure energy, since neither photons nor neutrinos have mass. (Some recent experiments suggest that neutrinos might have a *little* mass.)

One possible explanation, then, for the complete dominance of matter over antimatter on Earth would be that originally the amount of matter and antimatter in the universe was equal (as expected from symmetry considerations), but unequally distributed. In the vicinity of the the solar system, matter predominated. If, for example, you had 1 kilogram of matter and 0.9 kilogram of antimatter in close proximity to each other, you would very quickly have 0.1 kilogram of matter, no antimatter, and an explosion that would make a hydrogen bomb look like a popgun. If this scenario were true, then it would be most likely that the matter-antimatter unevenness would work itself out in patches the size of galaxies, rather than at the level of planets or stars. Thus, the Milky Way (our galaxy) would be expected to be all matter, but perhaps M31, the fairly near galaxy in Andromeda, would be all antimatter. This view can be sustained by the fact that cosmic rays detected on Earth seem almost always to be matter. Such cosmic rays are thought to originate in the Milky Way, suggesting but not proving that the Milky Way is all matter. As Dirac noted for stars, we cannot similarly learn much about M31. If any cosmic rays reach us from M31, they are few and far between and probably not identifiable as to origin. Our only source of information is the photon (the particle that, when thought of as a wave, is the basis of radio, infrared, optical, and x-ray astronomy). As noted, the photon is its own antiparticle. Thus, we cannot tell whether a photon reaching us from M31 or from some more distant galaxy comes from a region of matter or one of antimatter.

There is one possible way to tell. Neutrinos, unlike photons, come in two forms—antineutrios from normal neutron decay and neutrinos from antineutron decay. Like photons, neutrinos stream across intergalactic space in almost straight lines (very massive bodies near the path of a photon or neutrino can deflect it slightly according to Einstein's general theory of relativity; it was the confirmation of this effect by observation that made Einstein world famous). The problem is that neutrinos are difficult to detect. Several present-day neutrino "telescopes" are just barely able to detect neutrinos from the Sun. No neutrino telescope can separate the neutrinos from a specific galaxy and tell whether or not they were neutrinos or antineutrinos.

Despite the inability to tell whether or not there are large patches of antimatter somewhere in the universe, cosmologists generally believe that those large patches do not exist. Although the galaxies are now far apart, they were closer together in the past, since the universe has been expanding from the Big Bang onwards. Thus, in the past, galaxies would have interacted. If some had been antimatter galaxies, that interaction would have produced huge bursts of energetic photons—gamma rays. Energy does not just go away. If there were powerful gamma ray sources in the past, we would be able to detect them today. Since we do not, the conclusion is that the universe is *probably* predominantly made from matter.

Therefore, cosmologists need to explain why a process that is symmetrical on Earth today must, at the beginning of the universe, not been symmetrical. This is especially difficult since almost all laws of physics are symmetrical; indeed, until 1957, physicists would have said that symmetry existed to such an extent that no experiment could tell the difference between left and right. In 1957, however, results that will be discussed in Chapter 7 showed that this belief was not true. Further investigation along the same lines revealed that a particle, the K (kappa) meson, decayed in a way that was not symmetric for matter. Various different modes of decay for K mesons have turned out to be important tools in understanding particle physics. In one such mode, it can be shown that the K meson decays more often into a negative pion, a positron, and a neutrino than it does into their antiparticles, which are a positive pion, an electron, and an antineutrino. While this peculiar decay mode is unique, it shows that such a thing *can* happen in the real world.

Suppose, then, that at the Big Bang the first particles produced by this high energy included many that are more massive than any we find today in cosmic rays or that we can produce in our biggest particle accelerators. This is exactly what one would expect, since there was a lot of energy and by Einstein's formula the energy could take the form of massive particles. Suppose that one of these particles, call it X, had a decay mode that is unsymmetrical, like that of the K meson. The X particle decays more often to a proton than to an antiproton. Even if this effect is not very pronounced, even if it turns out that for every billion-and-one proton decays there were a billion antiproton decays, even if there were just this excess of one proton left after the other billion protons annihilated the billion antiprotons, even then, calculations show that there would be enough matter in the universe to account for what we see today.

As the universe cooled after the Big Bang, eventually there would not be enough energy to produce X particles. All the X particles would then decay, and we would be left with a universe filled with protons, where the only antiprotons seen are those created by high-energy reactions. This scenario assumes much that is far from proved. At present, it is still possible that the galaxy next door is made out of antimatter and that the hypothetical X particle never existed.

Furthermore, every particle has its antiparticle. When the Big Bang provided the energy to create X particles it would also have created equal numbers of anti-X particles. If the X particles had a tendency to decay into matter more than into antimatter, then the anti-X particles would have a similar tendency to decay into antimatter rather than into matter. The result is a Mexican standoff. We still should have the same amount of antimatter in the universe as matter.

The example of the K meson comes to mind, however. If the X particle were its own antiparticle (like the K meson) and if the X particle decayed into matter about a billion-and-one times for every billion times it decayed into matter, that would be enough. Theoretical physicists calculate that this is quite possible. In some new theories (the subject of Chapter 9 of this book), it is inevitable. Of course, when that happened, which would be very early in the course of the Big Bang—well before the first trillionth of a second had elapsed—all of the antimatter would annihilate as much of the matter as it could. In Dirac's suggestive imagery, each billion antiparticles would be a billion holes into which a billion particles would fall, leaving one lonely particle to carry on the good fight of the universe. As the particles fell, they would release energetic photons—gamma rays. The gamma rays, as noted before, would still be around, so we can check on these predictions. In 1965, Arno A. Penzias and Robert W. Wilson—who knew nothing of these speculations—were trying to eliminate all possible sources of noise in a microwave receiver (they worked for Bell Labs). At a wavelength of 7 centimeters, they persistently found a radiation level corresponding to a temperature of 3 degrees above absolute zero. It would not go away. What is more, this radiation seemed to come from everywhere in space with equal intensity. As luck would have it, a group of theoretical physicists were about to publish a paper explaining that the gamma rays caused by the process just described would have an energy of 3 degrees Celsius above absolute zero and be detected as the same in all directions in space. (For other reasons, such a radiation from the Big Bang had also been predicted by the originators of the Big Bang theory, but everyone had forgotten this prediction from the 1940s.) The experimental work of Penzias and Wilson was published alongside the theoretical work, convincing nearly everyone that the Big Bang really happened and winning the Nobel Prize for Penzias and Wilson.

Considering everything, one is tempted to say that it indeed all happened as described. The universe is all matter because the initial symmetry of matter and antimatter was broken in the first tiny fractions of a second after the Big Bang and then "frozen" into existence as the energy level cooled beyond the point where particles and antiparticles are created. Yet . . .

It is also tempting to think that the antigalaxies exist. What would life in an antigalaxy be like? Would time go backwards, for example?

At Play in the Fields of Physics

Sometimes an idea that starts simply as pure description gradually grows until it dominates our thinking. Democracy is an example. Certainly the initial idea came from the recognition that in a group of a dozen or so people more than half of them (if they agreed) could impose their will on the remaining less than half. Over the centuries, various implementations of this idea have been developed for larger and larger populations. Today, almost every organization on Earth *claims* that democracy is at least part of its government, even when, as with the stockholder democracy of modern corporations, the evidence for that claim is slight.

In physics something similar has happened with the idea of *field.* First of all, for those who have learned what a field is in mathematics, it is necessary to say that it is easier to understand a field in physics by forgetting about such mathematical fields as the field of real numbers or the field of complex numbers. While a physical field also behaves as a mathematical entity, the differences between the fields of physics and the fields of pure mathematics are greater than the similarities. As happens a lot in science and mathematics, only the names have been kept the same to confuse the innocent.

Since the concept of a field first became prominent in the nineteenth century, it is appropriate to start this discussion by looking at the work of a nineteenth-century painter who used a version of the same idea in his work—Georges Seurat. Seurat conceived of a mathematical way of painting. Each point on the two-dimensional canvas that was the space in which he worked was assigned a single, unmixed color. Seurat then applied these colors point by point. Up close, you could see that only the pure colors were used. From a slight distance, the eye merged the colors into subtle combinations. A color perceived, for example, as black might be formed from many dots of many different colors, none of which was black. Seurat used mathematical relationships he had developed for different colors to determine which color to use.

Modern color printing uses exactly the same method, although the dots are closer together. Use a magnifying glass to look closely at a *printed* color photograph and you will find dots that are colored magenta (a shade of red), cyan (a shade of blue), yellow, and black. The eye puts these together to create other colors; cyan and yellow dots look green without the magnifying glass, for example. Although printed pictures usually use four colors, including black, similar results can be obtained with just magenta, cyan, and yellow, although the dark colors are not quite so dark.

How is this related to the idea of a field? The most basic idea of a field is just the same as Seurat's method of painting. For a given space, such as the canvas, a quantity or a quality is assigned to each point. In Seurat's case, each point was assigned a color. The number of colors used can be related to the dimensions of the field, even though the painting is on a two-dimensional surface. If Seurat used just red, blue, and yellow, the painting would have three color dimensions superimposed on two spatial dimensions. Similarly, ordinary four-color printing could represent a four-dimensional field.

Another example of such a field would be the result of heating one edge of a metal plate. The heat would distribute itself on the plate so that points near the heat source would have high temperatures, while points further away would have lower temperatures. In this instance, the field would have numbers instead of colors assigned to each point. In both cases, the points could also be designated by numbers, specifically the numbers of a coordinate system.

Such a coordinate system could be assigned in various ways. You could, for example, use the bottom and left edges of the painting or the plate as the ordinary x and y axes. Or such axes might be positioned according to some physical part of the room, such as the intersection of two walls and one of the walls with the floor.

The assignment of the coordinate system does not change the relations between the colors in the painting or numbers assigned to the points on the plate. This is then a form of symmetry, for the relations between the numbers are invariant for such operations as rotation of the coordinate system and translation. In other words, if you hung a Seurat upside down or moved it to another place on the wall, the essential painting would not change. A similar rule applies to the example of the heated plate. For example, if the transformation were to change your method of measurement from a Celsius thermometer to a Fahrenheit thermometer, the actual numbers would change, but the relationship between them would not. You could still use the same formula to get from one number to another, and both numbers would be valid for measuring the temperature. Thus, switching to another thermometer is a symmetry operation for this kind of field.

But for some kinds of transformations, the whole painting would change. If the axes were no longer perpendicular to one another, or if the scale on the y axis was changed to double that on the x axis, or if you switched to polar coordinates, the relation between the colors on the painting that were assigned to various coordinates would change. This might then result in a different perception of color, since Seurat's mathematical relationships would be changed. Another kind of transformation, one that is more like those we will discuss when we get to particle fields, would be to "rotate" the field of the painting so that the red dots become blue, the blue dots become yellow, and the yellow dots become red. In printing a three-color picture, this could be accomplished simply by switching the plates or color separations, since each color is on a different printing plate. The result would still be a figure that we could recognize, although all the colors would be changed. Although the colors are changed, the shapes are invariant. Transformations that do not change the painting or plate are symmetrical; transformations that change the values are not. If a transformation leaves one characteristic of interest the same, such as the color substitution of the dots, that transformation is also considered symmetrical, even though other aspects of the field are changed.

These considerations lead to the modern definition of a field in physics: a field consists of an assignment of quantities to all points in space and the symmetry operations that leave those quantities unchanged. In general, physicists are concerned with much more complicated fields than the examples just given. For one thing, most fields that are interesting are defined in four-dimensional space-time. For another, the fields of particular interest have more complicated entities than colors or numbers assigned to each part, which changes the symmetry operations that are permitted in various ways.

While this is the modern view of a field, it did not arise all at once. It helps to look at earlier field concepts, especially since all of these ideas are still used routinely by modern physicists. The earlier ideas of what a field was all about emerged from studies of gravitation and the electromagnetic spectrum.

Suppose that you have a number of bodies in space. By Newtonian gravitational theory, each attracts each of the other ones with a force that depends on their two masses and the distance between them. Now consider placing a very light object (compared to the other bodies—a feather, say) somewhere in the midst of all this at a point we can call (x_1, y_1, z_1). Its motion will be directed in some specific direction by the combination of all the forces on it. Because it is very small compared to the other bodies, its motion will be a specified amount in a specified direction, and the amount of motion will define a specific force. That force and direction can then be assigned to the point (x_1, y_1, z_1). In such a manner, a unique force and direction can be assigned to each point in space. This assignment results in the Newtonian gravitational field that pervades three-dimensional space. The symmetry operations assigned to this field are those that give the field the same properties for an observer anywhere in space.

Note that if the feather were in a slightly different place—say a centimeter to the left or a centimeter down—its assigned force and direction would be different, but not very much different. If the difference in space were only half a centimeter, the assigned force and direction would be closer still to that of the original feather. Thus, it would be possible to draw a continuous line from one position to another showing the force and direction of the gravitational field. If the only bodies involved were two masses with one approximately double the mass of the other one (say that all other bodies are too far away to make much difference), then these lines would form a distinct pattern.

Figure 6–3 shows one of the conventions developed for showing fields, which is that the number of field lines drawn are

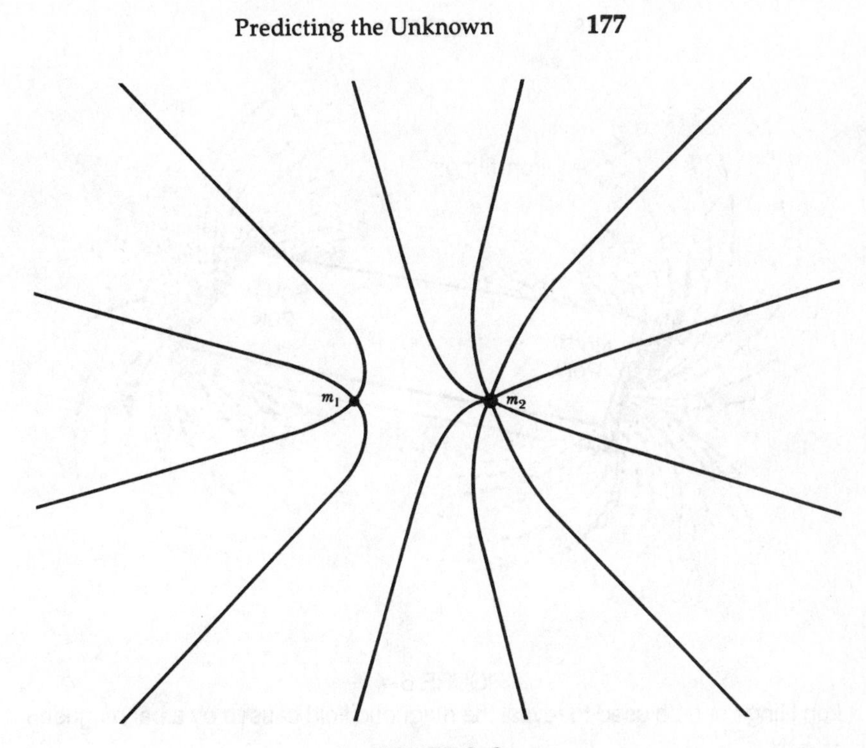

FIGURE 6–3
The gravitational field around masses m_1 and m_2.

proportional to the quantities being mapped. In this case, since the mass on the right is double the mass on the left, it gets 8 field lines, while the mass on the left gets 4 only. Another feature of the model is a result of the definition of field lines; the closer the lines are to each other, the stronger the field is (in this case, the higher the gravitational attraction on the feather). Notice that each field line for gravity has one end on a mass and the other extends infinitely out into space.

Figure 6–4 is another kind of field. Almost everyone is familiar with the experiment of putting iron filings on a sheet of cardboard that is lying on a magnet. The "field lines" of the magnet are displayed by the arrangement into which the iron filings fall, perhaps aided by a few taps on the cardboard to overcome friction.

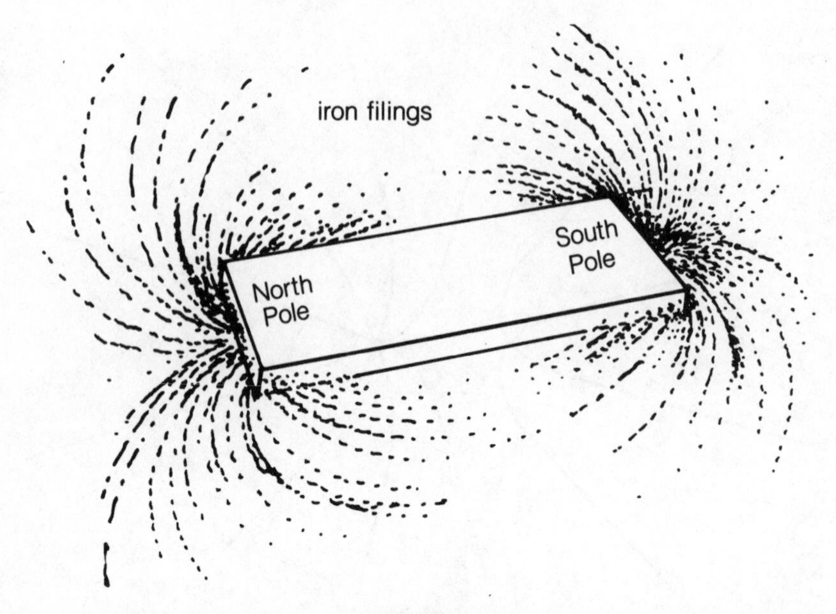

FIGURE 6–4
Iron filings can be used to reveal the magnetic field caused by a bar magnet.

Early nineteenth-century physicists, notably Michael Faraday, pictured these field lines as thin tubes connecting the north and south poles of a magnet. Similar field lines connect a source of negative electric charge to a source of positive electric charge. Faraday pictured these with the same tube model. The gravitational field can also be viewed as a set of actual field lines pervading space. Actual entities in space that form magnetic, electric, and gravitational fields eliminate the earlier concept of "action at a distance," which presumed that two magnets or planets attract each other through a vacuum that is actually empty. Faraday's vacuum was not empty. It was filled with field lines, so there was no need for action at a distance.

Newton's original concept of gravitation did use action at a distance, so he concentrated on particles and how these particles interacted. Switching to a field concept means putting the emphasis on the field. The interaction of the particles and the field determines how the particles will behave. Although the field is created by the particles, the field also controls the behavior of the particles, which behavior in turn changes the field. It is this rich connection between the cause of the field and the field itself that eventually came to dominate much of physics.

There is a significant difference between the gravitational field and the magnetic and electric fields. There is only one kind of mass, but there are two kinds of magnetic pole and two kinds of charge. Furthermore, opposite charges attract and likes repel, so that the electric fields formed by a positive and a negative charge or by two negative charges do not resemble the gravitational field at all. Instead, the fields formed by these combinations of charges look exactly like the fields formed by iron filings and magnets (Figure 6–5).

The electric field and the magnetic field not only resemble each other, but they are connected in a symmetrical way. If you move a magnetic field, an electric charge is produced. Similarly, if you move an electric field, a magnetic force appears. These fields are so intimately connected that nineteenth-century

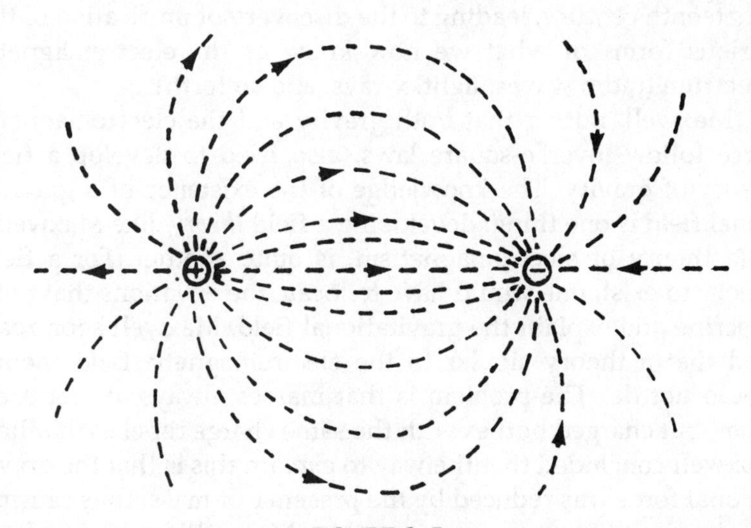

FIGURE 6–5

Like magnetic poles, a positive and a negative electric charge attract, while two charges of the same sign repel each other.

physicists, after the discovery of this connection in 1820, began to speak of the electromagnetic force. Before Faraday, however, physicists were dealing with three types of action at a distance: magnetism, electric charges, and the electromagnetic force.

James Clerk Maxwell was the first to take Faraday's ideas and put them into mathematical form. Like Faraday, Maxwell concluded that physicists were concentrating on the wrong part of the problem; they were looking at bodies and forces, but ignoring the fields. In 1865, Maxwell announced what he called "a theory of the *Electromagnetic Field,* because it has to do with the space in the neighborhood of the electric or magnetic bodies." One of his observations was that a moving charge would cause a magnetic force, which would also be moving, and that would then produce a moving charge, and so forth. Thus, the electric field and the magnetic field push each other, forming a wave.

By concentrating on the field, Maxwell was able to develop a set of 20 equations that described the electromagnetic field and its action. This was the most successful physical theory of the nineteenth century, leading to the discovery or unification of the various forms of what we now know as the electromagnetic spectrum (radio waves, light, x-rays, and so forth).

Maxwell, noting that both gravity and the electromagnetic force follow inverse-square laws, also tried to develop a field theory of gravity. The knowledge of the existence of a gravitational field is one thing; developing a field *theory,* like Maxwell's field theory of electromagnetism, is quite another. For a field theory to exist, one would have to locate the equations that both describe and explain the gravitational field. Maxwell soon realized that a theory similar to the electromagnetic field theory would not do. The problem is that masses always attract each other, but charged bodies with the same charge repel each other. Maxwell concluded that the way to explain this is that the gravitational force was reduced by the presence of mass, thus causing two bodies to want to move together. Maxwell wrote, "As I am unable to understand in what way a medium can possess such properties, I cannot go any further in this direction in searching for the cause of gravitation." For the next 50 years, all efforts to develop a field theory of gravity were unsuccessful.

One question regarding Maxwell's electromagnetic theory was whether or not the electromagnetic field required a medium (the ether) for its propagation. As we have seen, Einstein solved this question with the Special Theory of Relativity in 1905.

From the point of view of field theory, what the special theory does is to replace the symmetry transformations (the ones that do not change the field). Instead of simple translations and rotations, you must use the Fitzgerald-Lorentz transformation. This change led to a field that was invariant from the point of view of two observers who were moving steadily with respect to each other.

Einstein realized that he had not solved the complete problem. If one or both of the observers were being accelerated, the Fitzgerald-Lorentz transformation would not result in invariance. Over the next 10 years, he worked on this problem, developing in the process the first satisfactory field theory of gravity. The mathematics involved was much more complicated than anyone had suspected it might be. To this day, Einstein's field theory cannot be used to get exact solutions to even very simple problems, such as "how do two bodies attract each other?"

The Einstein field theory of gravitation is better known as the general theory of relativity. It appeared in its complete form in 1915, although Einstein published bits and pieces of it starting as early as 1907. The key to the general theory came when he realized that there existed frames of reference (that is, moving coordinate systems) for which the gravitational field was zero. Specifically, he noted that for "an observer falling from the roof of a house there exists—at least in his immediate surroundings—no gravitational field." Later Einstein called this insight "the happiest thought of my life." From this insight he eventually developed what he came to call "the principle of equivalence"—gravitational force and the force of inertia cannot be distinguished from each other.

Suppose you are in a closed room with no windows. Suddenly you feel that you are being drawn to the floor by a mysterious force. There are two equally likely explanations (in theory at least) as to what is causing the force: Something is accelerating the room through space (as in the beginning of an elevator trip to the top of a tall building); or the force of gravity has suddenly increased (as when your closed room rests on a very massive body, such as Jupiter). The principle of equivalence was also known to Isaac Newton, who discussed it in the *Principia,* but Einstein got a lot more out of it.

How did Einstein get from these two simple ideas to the field theory of gravitation? The route is rather more mathematical than the scope of this book, but we can look at a sort of simplified map of that route. First of all, the "happiest thought" means that measuring rods in free fall have a length that is not influenced by gravity, since gravity is exactly zero in a freely falling body. This may seem trivial until you recall that the length of a measuring rod as seen by two different observers moving uniformly with respect to each other is different for each of the observers. There is no preferred frame of reference. But for gravity, there is a frame of reference that can be used as a starting place, and it is free fall.

Second, the principle of equivalence suggests that an artificial gravity can be constructed by rotating a large ring in space. You may recall seeing various versions of a space station shaped like a giant turning wheel or spinning doughnut. These designs are based on the idea that such an artificial gravity would simplify life in space. As the wheel rotates, each point on the rim is constantly changing direction. A change in direction cannot be accomplished without acceleration (Newton's first law). As the wheel turns, points not at the center of the spacecraft are accelerated toward the center by one component of the circular motion. So the rim develops a fictitious force known as centrifugal force. One way to recognize that this force is fictitious is that the acceleration is the same on all bodies no matter what the mass is. Another fictitious force is the Coriolis force that drives winds and ocean currents on the surface of the Earth. Yet another fictitious force is gravity!

Why is gravity a fictitious force? Simon Stevinus and, later, Galileo showed that two objects of different mass dropped from a tower fell at the same rate. One reason that this result seems to fly in the face of common sense (as exemplified by the Aristotlean theory of motion) is that gravity behaves like a fictitious force instead of like a real force, which would have a different effect on different masses. One result of Einstein's field theory of gravitation was to show that gravity does not just behave "fictitiously," but it actually is fictitious. In fact, this should be expected from the principle of equivalence.

Picture life on one of these rotating space stations. Suppose that the station is actually an "Ark," a life-sustaining space ship containing a group of people escaping from some Earthly disaster, aimed for a distant star that might have a livable planet. The trip will take many generations. After the initial acceleration to put the Ark on its way out of the solar system, the ship is in free fall, but rotation provides "false" gravity around the rim. Thus, a gym at the center of the Ark is in free fall and provides the possibility of amazing acrobatics, while life at the rim is lived under ordinary gravity as far as anyone in the Ark can tell.

A mathematically minded passenger a few thousand years into the journey decides to map the Ark. He takes a meter stick and measures the diameter of the Ark. Since the stick is always at right angles to the acceleration, the length of the stick is the same as it would be in free fall in the gym at the center of the Ark. Then he measures the distance around rim in the same way with the same meter stick. Now, however, the meter stick is shorter than it is in free fall (because of the Fitzgerald-Lorentz contraction as established in the special theory of relativity). As a result, the circumference of the spinning circle is bigger than would be expected according to the formula that says it is π times the diameter of the circle.

Einstein realized that this means that Euclidean geometry is not valid on a spinning disk. But, by the principle of equivalence, Euclidean geometry must not be valid if you go from a region where there is no gravitational force (like the gym of the Ark) to one where there is a gravitational force (like the rim of the Ark). Another way to say this is that real space must be curved, not flat.

The gravitational field, in Einstein's theory, causes space to curve or warp. A common way to picture it is by thinking of a horizontal rubber sheet on which balls of various masses are placed. Each ball will distort the sheet by an amount that depends on its mass. In fact, each ball will be at the base of a valley. If two balls are sufficiently near each other, they will roll together, just as if a real force were pulling them together, not the fictitious force caused by the distortion of the rubber sheet. Thus, gravitational field theory or general relativity is also a theory of the geometry of space. Gravity no longer exists as a real force, but instead is the fictitious force caused by the space-warping property of mass.

Related to this is another line of thought that also stems from the principle of equivalence. The special theory of relativity leads to the conclusion that mass and energy are two aspects of the same thing. The principle of equivalence took this a step farther to say that energy, gravitational mass, and the apparent mass caused by inertia are all the same thing. A beam of light can be thought of as pure energy, but it still must obey gravity. Thus, it will be accelerated by gravity. Light from a star, fighting against the star's gravity, will develop longer wavelengths—be shifted to the red part of the spectrum. Light passing near a heavy body will be bent by gravity. The logical conclusion from this is that sufficiently intense gravity could prevent light from escaping at all—a black hole.

If this happens to light, then it also happens to any other material body that has a frequency. The frequency will be less in a gravitational field than in free fall. Therefore, identical clocks will go at different rates if they are placed in two different parts of a gravitational field, parts with a different force of gravity.

To define the gravitational field properly, one must take into account both the fact that gravity makes space curved and clocks run slow. A formula can be constructed that does this. It shows that mathematically there are 10 different gravitational forces. Four of them are related to the space-time coordinates x, y, z, and t. The other six are related to the various pairwise combinations of those coordinates: t with each of x, y, and z; and x with y, x

with z, and y with z. The first four describe (for a given system of masses moving in a given way) the energy (= mass) and momentum of the system. The other six describe the flow of momentum. The structure of space-time is determined by its energy (= mass) content and its momentum, so the structure of space-time and the contents of space-time are inseparable.

Here is where the mathematical difficulties get to be intense. Because the gravitational field is formed by moving masses, and because the masses are moving according to the gravitational field, one process feeds on the other. When the equations are found that describe this process, they turn out to be complicated types that are not easily solved.

Einstein, with the help of the work of mathematicians, was at least able to find the correct equations. Thus, he created the field theory that Maxwell despaired of finding.

Since 1915, there have been many experimental verifications of Einstein's gravitational field theory. The difficulty of the equations has meant that often only approximations to solutions are available for specific situations, so there are periodic challenges to Einstein's general theory of relativity, but so far no one has toppled it.

Maxwell's electromagnetic field theory and Einstein's gravitational field theory were major successes, but they did not resemble each other very much. It seemed to Einstein and to many other physicists that a single field theory ought to describe both fields. For much of the remainder of his life, Einstein worked—with very little success—on a "unified field theory."

One of those other physicists was Theodor Kaluza, a Polish physicist. Einstein had started with Maxwell's electromagnetic field, leading to special relativity, which in turn led to general relativity. Kaluza started with general relativity with a goal of obtaining the electromagnetic field. Astonishingly, he succeeded, but in such a bizarre way that his work was largely ignored for about 50 years after having somewhat of a vogue in the 1920s.

Theoretical physicists often succeed in this way: A hypothesis is offered that explains the unexplained, but which cannot be easily tested and leads nowhere. Such hypotheses litter the history of science. Sometimes changing knowledge will encourage others to lift the hypothesis from the trash heap years later, dust it off, and use it. More often the hypothesis is forgotten as other ideas come along—ideas that can be tested or generalized—to better explain the unexplained.

Kaluza may have been inspired by the success of Minkowski's version of special relativity in terms of a geometry of four dimensions, which was also adopted by Einstein in his essentially geometrical treatment of gravity. If adding another dimension to the three classical dimensions has produced such tremendous insights, why not go one step further and add a fifth dimension to the four of Einstein's space-time? Kaluza tried it and it worked. If space had five dimensions, the gravitational field would behave in the fifth dimension exactly like the electromagnetic field. In five dimensions, there was only one field. Kaluza had created the second unified field theory. (The first unified field theory was developed by Hermann Weyl for four-dimensions. Although it was deemed a clever failure by Einstein, Weyl invented gauge transformations in the process; which, as we will see later, figure mightily in more recent field theories.)

The basic idea behind Kaluza's theories is the usual combination of symmetry and invariance. A field is described in terms of its symmetry; specifically, length squared and a certain mathematical form for the fifth dimension are invariant for transformations of the field. The fifth dimension behaves like a spatial dimension, not like time. The field resembles a cylinder because of the way the mathematical form for the fifth dimension is kept invariant. Then it is possible to show that the combined gravitational-magnetic field warps this cylinder universe just the way that the gravitational field warps the four-dimensional universe. Charged particles in the warped space move in the equivalent of straight lines (known to scientists as geodesics).

Einstein was impressed. Even before Kaluza's paper on the subject was published (in 1921), Einstein knew of the work and had written to Kaluza saying "At first glance, I like your idea enormously." Nevertheless, physicists were bothered that the fifth dimension was spacelike, but not observed. In 1926, however, Oskar Klein reworked and extended Kaluza's theory, showing that we did not perceive the fifth dimension because it was too small. After Klein's work, theories that the universe has more than four dimensions tended to be based on similar ideas and are known as Kaluza-Klein theories.

What does it mean to say that the fifth dimension is too small to be observed? The symmetry conditions used in the development of the field resulted in a space that was like a cylinder. Klein showed that this was literally so; the fifth dimension could be thought of as a creating an infinite tube. It is the size of this tube that results in our inability to detect the fifth dimension. Such a dimension corresponds to the mathematical condition known as *compactification.* A Kaluza-Klein theory is one that has the familiar four dimensions of space-time and one or more additional compact dimensions.

It helps to think of converting a one-dimensional universe to a universe with a compact second dimension. At every point on a one-dimensional line, a particle can move in two directions. Now add to each point the possibility that the particle could also move in a loop, so long as such a motion always brought it back to the original point. This loop would be the second dimension. The path a particle might take from one point on the line to another might include any number of trips through these loops as well. Thus, people picture a space with one compact dimension as a sort of tube. If you were going from four dimensions to five, the compact dimension would still be a loop at each point, so a path of a particle in space-time that appears like a line from a four-dimensional perspective, is actually a tube instead of a line.

Klein discovered that the diameter of this tube must be incredibly small—0.8×10^{-30} cm, or the digit 8 preceded by 29 zeros and a decimal point. Since this is very much smaller than the nucleus of an atom, it is no wonder that we do not observe the fifth dimension.

Still, it was not clear whether the fifth dimension was a mathematical fiction that worked or an actual dimension. As noted at the beginning of this chapter, mathematical descriptions that work are often taken to be reality. In this case, however, the idea of a five-dimensional world may have seemed too strange for many physicists, although Einstein and others returned to it from time to time throughout the 1930s. The fifth dimension worked, but it did not lead anywhere—no new discoveries came from this theory. Eventually, most physicists ignored it.

Then, in the 1980s, a number of ideas came together again in previously unsuspected ways. Field theories and Kaluza-Klein theories became hot news again—but now it seemed as if there should be 10 or 11 dimensions, not 5. The story of these developments, however, is best left until Chapter 9 when there will be additional background to help understand them.

7

Of Time and the Mirror

Why is nature so nearly symmetrical? No one has any idea why. The only thing we might suggest is something like this: There is a gate in Japan, a gate in Neiko, which is sometimes called by the Japanese the most beautiful gate in all Japan; it was built in a time when there was great influence from Chinese art. This gate is very elaborate, with lots of gables and beautiful carvings and lots of columns and dragon heads and princes carved into pillars, and so on. But when one looks closely he sees that in the elaborate and complex design along one of the pillars, one of the small design elements is carved upside down; otherwise the thing is completely symmetrical. If one asks why this is, the story is that it was carved upside down so that the gods will not be jealous of the perfection of man. So they purposely put the error in there, so that the gods would not be jealous and get angry with human beings.

We might like to turn the idea around and think that the true explanation of the near symmetry of nature is this: that God made the laws only nearly symmetrical so that we should not be jealous of His perfection.

Richard P. Feynman

"I don't understand you," said Alice. "It's dreadfully confusing!"

"That's the effect of living backwards," the Queen said kindly: "it always makes one a little giddy at first—"

"Living backwards!" Alice repeated in great astonishment. "I never heard of such a thing!"

"—but there's one great advantage in it, that one's memory works both ways."

"I'm sure mine only works one way," Alice remarked. "I can't remember things before they happen."

"It's a poor sort of memory that only works backwards," the Queen remarked.

Lewis Carroll

An accountant worked many years at the same firm. Every morning his colleagues observed that as he began work he unlocked and opened the top drawer of his desk, stared intently into it, shut and locked it, and began work. Gossip produced many theories about what could be in that drawer, because no one was ever allowed to see its contents. Could it be a photograph of some long lost love? A magic talisman of some kind?

Finally, the accountant retired. At his retirement party, one colleague, emboldened by drink, asked the old accountant about the mysterious drawer. After a moment's thought, the old accountant murmured, "I think it won't hurt now if people know," walked to the desk, pulled out the key, and unlocked the drawer. A small crowd gathered to peer into the drawer.

All that was there was a short note, reading "The credits are on the left and the debits are on the right."

Right or Left?

It is generally assumed by people who hear this story that the accountant knew which was right and which was left, but could not recall where the debits and credits went. I prefer to think that he could tell a debit from a credit, but had trouble remembering which was right and which was left.

Very many people—I am one of them—have difficulty remembering which is right and which is left. No one not in free fall has a problem distinguishing top from bottom. Gravity provides a ready reference point. In a freely falling spacecraft, it might be necessary to leave a note saying "The insignia is the top side of the room" or something of the sort, but in the vicinity of a planet, something seems to want to accelerate your mass toward its center, so down is clearly down and up is clearly up. We know that the tendency to move toward the planet's center is caused by the mass of the planet warping space-time rather more than your own mass warps space-time; in other words, gravity determines "down" near a large mass.

Similarly, another pair of directions is uniquely defined by our sense of self and nonself. An object is either in front of us or in back of us. Would a perfectly symmetrical creature, sort of an octopus with eyes all around it and a mouth at the top of its head, have a similar perception? Such a creature—say, an alien being on another planet—might be forced to define front and back in another way, but for a given object, the alien should be able to identify the part of the object that is near and the part that is far, or equivalently the front and back. Thus, gravity and a sense of self–nonself can fix two of the four dimensions. This is not to say that a universal coordinate system is defined, just that one being can communicate with another about those parts of a coordinate system that are defined in relationship to itself.

A third set of coordinates seems to be uniquely defined for everyone but the White Queen from *Through the Looking Glass.* Like Alice, we have "a poor sort of memory that only works backward." While philosophers have written long, tedious books defining the nature of time, the rest of us have relied upon our poor sort of memory to know which way time's arrow points. A few individuals, injured in surgery or by cardiovascular accident, lose their memory and live always in the present, but the rest of us can easily discern yesterday and recognize the direction of tomorrow. We assume—although we do not know—that the imaginary alien on the other planet would also have a similar sense of time.

Later in the chapter, we will see that it is not all that easy—the philosophers were right to wonder about the nature of time— but for our purposes, just as we have up-down and front-back to rely on, we can also utilize past-future.

This leaves one set of coordinates unassigned. How can we make a similar distinction between right and left? Do we, like the accountant, need to keep a note in the drawer? As I noted, I have a mild problem in telling right from left. When suddenly in doubt as to which is which, I mentally picture a page from a book. I remember that I read from left to right. Therefore, the start of a line of type in the book is on the left, while its end is on the right. Young children who are learning to read and who have right-left dyslexia need other crutches. A teacher may tie a piece of cloth around the child's left hand to call attention to the side of the page where reading is to start. Adults use different crutches, a watch or a ring for example. The debits are on the left and the credits are on the right. Or, in another version of the tale, "Port is left and starboard right."

A philosophical question is whether or not the fourth dimension (left-right) can be defined uniquely without some equivalent of a note or a cloth tied around the arm. That is, given a creature isolated physically, our alien on another planet, say, who can easily determine up-down, front-back, and past-future with the same meanings as you or I would assign, can the alien also determine left-right in such a way that it will mean the same as it does to us on Earth?

Questions such as this have been troubling philosophers since at least Immanuel Kant in the eighteenth century. Kant imagined a human hand as the only object in space. How could you tell whether it is a right hand or a left hand with no other object for reference? A hundred years later, William James put the matter in a framework closer to the one we have been using. Suppose there is a cube. You want to tell someone who does not know your coordinate system how to discern the three space dimensions by painting the faces of the cube in six different colors. Gravity and self–nonself easily provide ways to paint the top, bottom, front, and back. But how do you communicate the idea that the left side should be painted red and the right side blue?

This may not seem like a very practical problem, but it does have its practical side. Since two coordinate systems are possible, physicists and engineers have to have a way to tell them apart. One such case occurs in a mathematical operation called *vector multiplication.* A vector in three-space or four-space can be conveniently thought of as an arrow of a given length pointing in a given direction. If you have two such vectors, a third vector, called the *vector product,* can be uniquely defined as an arrow of a certain length that is perpendicular to the plane defined by the first two. But for the vector product to be unique, it is necessary to say which way it points. If the plane of the first two is, say, horizontal, does the product vector point up or down? A bridge that was built with product vectors randomly assigned might not be all that safe. Fortunately, most people, including engineers and mathematicians, do know their right hand from their left. Consequently, they have relied on what is known as "the right-hand rule" to keep all their product vectors (and coordinate systems and even electrical currents) pointing in the correct direction.

The right-hand rule is almost never written down; instead it is communicated verbally from teacher to student. Although the right-hand rule is not completely an oral tradition, most textbooks rely on a picture to show which system they have in mind, while some use an equivalent, but less handy rule about right-handed screws. Simply put, however, the right-hand rule is that *if the thumb of your right hand points along the positive direction of the first vector and the index finger along the positive direction of the second vector, then your remaining fingers, if bent at a 90 degree angle toward your palm, will point at the positive direction of the vector product.* Since the vectors are typically shown so that the first vector points directly out of the page at you and the second vector points to the right, to verify that a vector product actually obeys the right-hand rule requires a bit of the contortionist's skill (Figure 7–1).

The other mnemonic for telling a coordinate system's or vector product's handedness depends on knowing what a right-handed screw is. A right-handed screw will move into the wood when it is turned clockwise. If you rotate the first vector into the

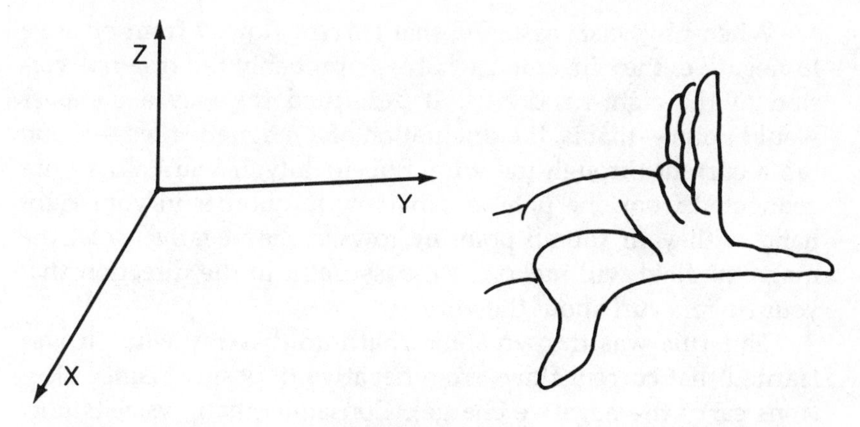

FIGURE 7–1
One way to remember how a vector product works is to use the right-hand rule.

second in such a way that it is clockwise, then the third vector will represent the body of the screw. The right-handed-screw rule requires more mental gymnastics than physical ones to apply, but gives the same result.

Neither of these rules would suffice to tell our imaginary alien which was right and which was left. The alien was not even born with hands, much less a right one and a left one. Also, unless one could physically transmit a screw or some electronic equivalent through space, there would be no way to apply the right-hand-screw rule.

Perhaps an "electronic equivalent" might supply the key to this dilemma. After all, if there were some way to transmit a television picture from us to the alien, you could put a square on the left-hand side of the screen, which would be the equivalent of tying a cloth around the left hand. Unfortunately, it would be just as easy to make a television set that showed that square on the right as on the left. The components of the set would all be reversed and the current would flow in the other direction. You cannot communicate the direction of the current, because it too depends on the difference between right and left.

When physicists assumed that current flowed from positive to negative, they invented what was probably the original version of the right-hand rule. It described the way a compass would point—that is, the orientation of a magnetic field—if you ran a current through the wire. Specifically, if you hold a wire connected from one pole of a battery to another in your right hand with your thumb pointing toward the negative pole, the magnetic field will make a compass point in the direction that your fingers curl about the wire.

This rule was thrown somewhat into disarray when it was learned that current flows from negative to positive (since electrons carry the negative charge). Consequently, physicists now point the thumbs of their left hands toward the positive pole, and their fingers point in the correct direction. In this case, the left-hand rule produces the same result as the right-hand rule because one points the thumb at the negative pole and the other points the thumb at the positive pole.

Thus, without being able to communicate to the alien which way the current was flowing, you still could not communicate left from right by a television picture or any other electronic means.

One further possibility might occur to you, however. Since the current flow determines the way that a compass points, by symmetry the way that a compass points could determine which way the current flows. That could only work, however, if you could communicate which pole of the compass is north and which is south. On Earth, the problem seems easy. The pole of the compass that points to the Earth's north pole is the south pole of the compass (people often forget that a compass's "north pole" should really be called its "north-seeking pole"). The alien would have no way to establish which was the north pole of its planet. Therefore, there would be no practical way of telling which way the compass was pointing, and no way to determine which way the current was flowing, and, finally, no way to build a television set for which you could be sure that a square transmitted to be on the left side of the screen actually appeared on the left.

Recently, scientists faced real problems of this type. For seven of the nine known planets in the solar system, it is fairly easy to determine which is the north pole and which is the south. Since those seven planets rotate in the same plane as the Earth (roughly), we can assume that if they have a magnetic field, the north pole is on the same side of the plane as Earth's. Venus is a problem, however, since it rotates retrograde; that is, if an observer were outside the plane determined by the planetary orbits, Venus would appear to be slowly turning clockwise, whereas Mercury, Earth, Mars, Jupiter, Saturn, Neptune, and (probably) Pluto would appear to be turning counterclockwise. Astronomers generally conclude that Venus is turning backwards, but of course it is possible that Venus is upside down. To determine this, one would need to know which is the north and which is the south pole. One way to do that would be to measure the planet's magnetic field, which cannot be done from Earth but which has been measured from space. As it happens, the Venusian magnetic field is very weak, possibly as a result of its slow rotation—the Venusian day is 18 days *longer* than its year—and these measurements have not yielded a definite answer to the question. Still, most astronomers prefer to think that Venus is backwards instead of topsy-turvy, so maps of Venus show north in the conventional direction.

The other problem is Uranus. It rotates on its side, and its rotation, like that of Venus, is also retrograde. From Earth, we sometimes see a pole facing us, which means that the rotation is seen sometimes as clockwise or counterclockwise. When the poles do not face Earth, the rotation appears as up-and-down motion, like a barrel rolling. Every 42 years, we are viewing a different pole, but from Earth there is no way to tell which one. When Voyager 2 traveled to Uranus in 1986, it was generally thought that the south pole was facing Earth. Determination of the magnetic field was of major interest on this mission (not really for the purposes in mind in this discussion), so that should have settled it. It did to some degree, and the astronomers were wrong. We were looking at the north pole. However, the north magnetic pole is not near the true north pole, which is what we would expect from our experience with other planets. It is tipped by 55 degrees. On Earth that would be like having the north magnetic pole in Minneapolis.

It would appear that the purpose of this digression into astronomy is to note that the only way to tell right from left (or north from south or clockwise from counterclockwise, which turn out to be equivalent problems) to someone on another planet or a distant star would appear to be to transport a physical object to the planet. The planet must be distant, for otherwise we could communicate clockwise and counterclockwise by having the alien view the rotating Earth. Since all these problems are equivalent, solving one of them by observation would lead to the solution of the other two.

Actually, the problem has been solved without the necessity of communicating a physical object. It has been known for over 30 years that there is a way to communicate pure information about right and left that resolves these problems. The method, however, is so surprising that it deserves a section to itself.

A Particle with Hair?

In the beginning, the problem was not right or left, it was one or two, or, put another way, even or odd.

In quantum theories of particles, a particle is completely determined by a set of numbers that describe it, as was noted at the end of Chapter 4. These numbers are the quantum numbers of a particle, and they are not especially mysterious. The idea is that if you know the mass, charge (expressed as a number, charge is +1, 0, or −1), spin, magnetic moment, and so forth, you know what the particle is. What quantum numbers have in common is that they are integral multiples of some fixed amount; they are not just any real number that comes along.

Although we think of quantum numbers as applying to very tiny particles only, something very similar exists for the largest concentrated masses—black holes. Only three numbers are needed to describe a black hole—its mass, its amount of rotation, and its electric charge. Any two black holes that have the same three numbers are *exactly* alike. Astronomers put it this way: "A black hole has no hair." Thus, since a set of quantum numbers defines a particle completely, a subatomic particle also has no hair.

Quantum numbers originated with the Bohr model of the atom in which a number for the size of the orbit of an electron, a number for the shape of the orbit, a number for the magnetic moment of the electron, and a number for the electron's spin were found to describe completely an electron in an atom. In 1928, Dirac developed symmetric equations for the electron that specified charge, spin, mass, magnetic moment, and wave properties for free electrons, a set of quantum numbers that seemed at first to be the end of the road, but actually was only the beginning. At that time, only the electron, proton, and the photon were known. Hints of other particles were in the air: Pauli had postulated the neutrino the previous year, and Dirac's own equation implied the positron, although that was not understood at the time.

Where do these quantum numbers come from? They are determined by symmetry considerations. Noether's theorem tells us that each form of symmetry is intimately related to a specific conservation law. If the mass is conserved, or the spin, or the magnetic moment, or the charge, then there is a specific symmetry operation that is invariant. An example involves Pauli's prediction of the existence of the neutrino. Although it was still five years before the discovery of the neutron, physicists had observed that in the nucleus of an atom, it sometimes happened that an electron was produced and at the same time a proton appeared (today we know that a neutron has decayed, producing an electron, a proton, and a neutrino, but it was thought then that the electron and proton were both present initially and their charges cancelled, effectively hiding the electron). This event, which is called *beta decay* because the emitted electrons are called beta rays, conserves charge: A negatively charged electron and a positively charged proton appear at the same time. Mass-energy should also be conserved—Einstein had already shown that mass and energy are two manifestations of the same thing. Careful measurements showed, however, that some energy is lost in beta decay. To restore the symmetry—that is, the conservation law—Pauli declared that a third particle, one without charge, must be present, accounting for the lost energy. Although it would be nearly 30 years before such a particle was detected, the particle could be described in terms of its quantum numbers. We now know that the predicted particle was the antineutrino, although its original name, suggested by Enrico Fermi, was *neutrino,* Italian for "little neutral one."

The discovery of the neutron and positron led to further refinements in the idea of quantum numbers, as well as to a reasonable expectation that there might be more undiscovered particles around. In particular, the discovery of the neutron, which seemed so much like the proton except for charge, a slight difference in mass, and the habit of beta decay when left to itself, led both to new particles and a new quantum number.

The new particles emerged from studies of what could hold the neutrons and protons together in the nucleus of an atom. A hydrogen atom has one proton most of the time, but it sometimes has a proton and a neutron or even a proton and two neutrons. All other atoms have nuclei that contain both protons and neutrons. Now the problem is that, except for the hydrogen atoms, this behavior could not be explained on the basis of forces known in 1932 when the neutron was discovered, which were gravity and the electromagnetic force. Gravity is much too weak an attractive force to overcome the strong repulsive force of the positive protons. Fermi tackled the problem and developed a beautiful theory that predicted an attractive force, but the force turned out to be also too weak, even when gravity was thrown in on its side, to overcome the repulsive electromagnetic force. Eventually, however, this weak force became central to other new developments, keeping as its name the *weak force* or the *weak interaction.*

Hideki Yukawa then worked on the problem; there must be a strong force. A good theory had been developed to describe the electromagnetic force in terms of the exchange of particles. Fermi's theory also had used exchange of particles to predict a force. Yukawa predicted that there would be a new particle found that had the right set of quantum numbers to account for the strong force between protons and neutrons. Such a particle, when exchanged, would turn a neutron into a proton, or vice-versa. The law of conservation of charge showed that the charge of the predicted particle would have to be either +1 or −1. From the measured amount of the strong force, it was possible to predict that the mass would be about 200 times as great as the electron. Finally, since neutrons and protons both have a spin of 1/2, the spin of the predicted particle must be 0 (a result that also arises from other considerations). With the quantum numbers in hand, one could keep an eye out for the predicted particle.

What happened first was a false alarm. In 1937, two years after Yukawa's article was published, a particle with the right mass was discovered and was assumed to be Yukawa's particle. Closer examination showed that the new particle, now known as the *muon,* was no good. Not only was it weakly interacting instead of strongly interacting, but it had the wrong spin, 1/2 instead of 0. Ten years later, however, the *pion* was found to be the predicted particle. Robert Oppenheimer later referred to this period in particle physics as a "ten-year joke," but it should be noted that for most of those ten years, World War II focused physicists more on practical problems than theoretical ones.

From the end of the 1930s through the end of the 1940s, particle physics was almost orderly. The predictions of Pauli, Dirac, and Yukawa seemed to have accounted for all the known particles even before the actual particles were found. The neutron had been an unexpected surprise, but it fit well into the theories of the time. Only the muon, still thought to be Yukawa's particle for most of this time, did not fit. The problem with the muon is that it had all the same quantum numbers as the electron except for mass (it is 207 times the mass of the electron), and no one could explain why. In fact, physicists today are still uncertain about where the muon fits into their current picture of the universe, often quoting I. I. Rabi on the subject: "Who ordered *that?*"

During the period from 1946 through 1953, however, all this apparent order was thrown into chaos. At that time, before really powerful particle accelerators had been built, the most energetic processes available to particle physicists were cosmic rays. By sending photographic films into the upper atmosphere on balloons or at the tops of mountains, physicists could obtain a record of cosmic-ray interactions. When the films were developed, however, strangers appeared. These strangers were particles that not only were not predicted, but that misbehaved. According to the existing theories, the new particles should decay into other, more familiar particles almost instantly—in about 0.000000000000000000000001 second. But the tracks on the photographic plates showed decay times a hundred trillion times longer than that. Something was keeping the particles from decaying.

Soon a reasonable theory was proposed to explain these strange particles. If they did not decay sufficiently fast, there must be something conserved, yet another symmetry to be found, and therefore a previously unknown quantum number. At first, a new quantum number introduced by Heisenberg, called *isotopic spin,* was revived as the likely quantity to be conserved, but that still left some mysteries. In 1953, Murray Gell-Mann (and independently Kazuhiko Nishijima) introduced yet another quantum number that did the trick. Since little was known about this new quantum number except that strange particles had it, Gell-Mann named it *strangeness.* Strangeness led to great breakthroughs in understanding particle physics by the methods of symmetry—the subject of Chapter 8.

Why does strangeness slow the decay of strange particles? Other conservation laws *prevent* particles from decaying. For example, the electron cannot decay because of conservation of charge; there is no particle with a negative charge that has less mass than an electron, so there is nothing for an electron to decay into. On the other hand, the muon, which has the same charge as an electron but has 207 times the mass, quickly decays into an electron, an antineutrino, and a neutrino.

The strange thing about strangeness is that it is only conserved some of the time. Specifically, it is conserved in strong interactions (mediated by Yukawa's pion force), but it is not conserved in weak interactions (mediated by the weak force).

When a strange particle is first formed, it cannot decay by strong interactions because of strangeness. But it can decay by a weak interaction, which it does. This process often reduces the strangeness of the particles in the decay products—technically, it increases the number, since strangeness is assigned negative quantum numbers. Thus, the strangeness of a particle such as the lambda meson (the first strange particle to be found) is −1, while the strangeness of a proton or a neutron is 0. When the lambda decays, it produces either a proton or a neutron along with other particles that have no strangeness assigned. The quantum number is increased by 1 and the strangeness is decreased to 0. A particle with a strangeness of −2 is stranger than a lambda particle, so it first decays to a lambda of strangeness −1, which then decays to a particle of strangeness 0. Depending on which strange particle you start with, similar chains may be formed, eventually resulting in a group of decay products that have no strangeness at all, and therefore behave like—and in fact, are—familiar particles. Since the weak interactions are much slower than the strong ones, the whole decay process is slowed down. Thus, strangeness accounted for itself.

One group of strange particles discovered around 1950 were called the K mesons. (The K was originally the Greek letter kappa, which unfortunately looks exactly like the Roman letter we call "kay." Later K, or kappa, mesons came to be called *kaons,* which suggests to me that most physicists called them "kay mesons" instead of "kappa mesons" from early on.) While conservation of strangeness accounted for the length of time it took the K mesons to decay, closer study of these strange particles revealed something that could not even be explained by strangeness. Two K mesons that had all the same quantum numbers, meaning that they should be the same particle, could decay in two different ways. One of them, called the tau particle, decayed into three pions. The other one, called the theta, decayed into two pions. The only way you could tell a tau from a theta was after the fact, since by all possible means of identification before they decayed, the tau and the theta appeared to be the same particle. It appeared that K mesons had hair. This new problem was even more unsettling than the original discovery of strange particles. It came to be known as the *tau-theta paradox.*

Note that there is a lot of confusion about names involved here. After the tau-theta paradox was resolved, the names *tau* and *theta* were dropped. Later, in 1975, a particle similar to the muon was discovered, and named *tau,* even though it had nothing to do with the *tau* meson of the mid-1950s. Some physicists call the more recently discovered particle the *tauon,* in parallel with the muon and pion, which helps a little. Today the tau meson from the 1950s is known simply as the kaon. Further confusion results when the *tauon* is referred to as a "tau meson;" for whatever the *tauon* is, it is certainly not a meson. In the discussion that follows, the names *tau* and *theta* will be used with the meanings they had in the mid-1950s.

As you will recall from Chapter 4, all subatomic particles have an even or odd parity assigned to them, based on their spin. Particles also have a plus or a minus parity, which will be discussed here. It is the plus-or-minus parity that caused the tau-theta crisis.

Plus-or-minus parity, like even-odd parity, is related to the angular momentum of a particle, but not the component of angular momentum that is described as spin. Another component of angular momentum is called *orbital angular momentum,* and it is also described by the wave function of the particle. As in the case of even-odd parity, if this wave function is unchanged by mirror reflection, the wave function is even, but it is said that the particle has a parity of +1. Similarly, if the wave function switches left to right upon mirror reflection, the wave function is odd, and the particle has a parity of −1. These values for the parity are chosen because particle interactions, ranging from the emission of a photon by an electron in orbit in an atom to the decay of a pion into a muon and a neutrino, obey the rules of multiplication for signed numbers. In the former case, that of the electron emitting the photon, the emitted photon has a parity of −1. To keep the total parity of the system the same, the electron in orbit changes parity by moving to a different orbital level. If the electron (before emission) had a parity of +1, the emission of the photon would shift the electron to -1, so that the total state of the electron and photon considered together, -1×-1, would continue to be $+1$. If the electron originally had a state of -1, the electron would switch to $+1$, so that the total state would remain $-1 \times +1 = -1$. Similarly, in a decay reaction, where the parities of the particles are intrinsic, not determined by orbital levels, the total state of the decay products must be equal to the state of the original particle. When a pion decays into a muon and a neutrino, the determining factor is that the pion has a parity of -1. Since there are two particles after decay, one of them must have a parity of $+1$ and the other of -1, since $+1 \times -1 = -1 \times +1 = -1$.

Now we come to the essence of the tau-theta paradox. The form of the kaon that was called tau decayed into three pions. Since a pion has a parity of -1, the kaon must have a parity of -1, since $-1 \times -1 \times -1 = -1$. But the form of the kaon that was called theta decayed into two pions, which implies a parity of $+1$, since $-1 \times -1 = +1$. But the only way you could tell a tau from a theta was after it had decayed. It seemed impossible that otherwise identical particles could have different parities.

The Lord Is a Weak Left-Hander

Particle physicists were searching for a way out of the tau-theta paradox. Although this was seemingly accomplished dramatically early in 1957, like most advances in science it was accomplished over a period of time and various people contributed to the solution.

At first, many physicists chose to hope that somehow it could be shown that tau and theta were really different particles; or, alternatively, that tau and theta were the same particle, but that there was some reasonable explanation, based on known physical laws, to account for this particle's two conflicting decay modes. The Australian physicist R. H. Dalitz undertook a careful study of this second possibility in 1954 and 1955 and effectively ruled it out according to the accepted physical laws of the time.

But if tau and theta were different particles, why were they exactly alike until they decayed?

Most of the physicists interested in this problem came to the University of Rochester in upstate New York in April 1956 for the Sixth Annual Rochester Conference on High Energy Nuclear Physics. Among these were experimentalist Martin Block and theoretician Richard Feynman, who shared a room. Block and Feynman discussed the problem every night for a week before a closing session devoted to strange particles, of which tau and theta were the strangest, in the ordinary sense of the word at least. Block earned a place in the history of physics by suggesting to Feynman that perhaps the law of conservation of parity did not work all the time. Feynman thought it was an interesting idea, since he did not know of any experimental evidence that conclusively showed that this conservation law really did hold all the time. It just seemed so obvious that the law should hold that physicists had taken it for granted. Recall, however, that physicists had just learned that strangeness was conserved in strong interactions, but not in weak ones. Thus, there was already a precedent for a conservation law that was less than universal.

Feynman did not really believe that parity was not conserved, however, because, as he pointed out to Block, that would mean that you could tell the difference between left and right. As discussed in the first part of this chapter, assaults on the question of telling left from right had all resulted in defeat. Because conservation of parity was the equivalent of not being able to tell left from right was one of the main reasons that physicists took it for granted.

All conservation laws, of course, have a basis in one kind of symmetry or another. It had been shown that the symmetry behind parity conservation was right-left invariance. Another way of saying this is that if a particular reaction among particles occurred, then its mirror image would also occur with equal probability. This idea was much older than the discovery of parity conservation; it was involved, for example, in the problem of tartaric and racemic acid that Pasteur investigated. As far as anyone could see at that time, right-left invariance should be true of all physical reactions.

We know that Feynman did not believe that right-left invariance could be overthrown, because he later bet physicist Norman Ramsey 50 dollars that experimental evidence would show that parity was conserved. Nevertheless, Feynman was intrigued by Block's idea, so the last day of the conference, after talks on the tau-theta paradox and other aspects of strange particles by Chen Ning Yang (known to colleagues as Frank Yang) and Murray Gell-Mann, Feynman rose and said, "I am asking this question for Martin Block," and proceeded to ask if it was possible that parity was not conserved in weak interactions.

Yang was already heavily involved in the tau-theta paradox. He was the one who established that the parity of the pion was −1, setting the stage for the problem in the first place. He and Columbia University physicist Tsung-Dao Lee had worked comfortably together on other projects, and it was natural for them to work together on the tau-theta paradox after that April meeting in Rochester. In a famous conversation in early May of 1956 at the White Rose Cafe near Columbia University, they decided to examine jointly all known experiments involving weak interactions to see whether or not parity conservation could be determined from the known experimental record.

It should be noted that both Yang and Lee are theoreticians, not experimentalists. Yang as a graduate student had an effect on laboratory work similar to the famous Pauli effect. His fellow students put it this way: "Where there's a bang, there's a Yang."

Late in June 1956, Yang and Lee put their findings and suggestions for new experiments into a paper called "Question of Parity Conservation in Weak Interactions." In it, they pointed out that the tau-theta paradox could be taken as evidence of nonconservation of parity in weak interactions, but that not enough was known about strange particles to be sure. Later Yang described their feelings at that time in this way: "The fact that parity conservation in the weak interactions was believed for so long without experimental support was very startling. But what was more startling was the prospect that a space-time symmetry law that the physicists had learned so well could be violated. This prospect did not appeal to us." In their paper, Yang and Lee gave a detailed list of experiments that could be done with familiar particles and interactions to test parity conservation. Experiments along those lines were soon started by a joint group from Columbia University and the National Bureau of Standards, as well as by a different group at the University of Chicago.

The first semipublic announcement of results came at a "Chinese Lunch" at Columbia University, a regular Friday entertainment of the Columbia physics department, on January 4, 1957. Lee had been the instigator of the Chinese Lunches and tended to take charge of ordering for everyone. The Columbia representative of the Columbia-Bureau of Standards experiment, Chien-Shiung Wu, was not present at this lunch, but she had been keeping Lee informed about the progress of the parity experiments. Lee was able to report that parity was not conserved in weak interactions.

Measuring parity itself is rather difficult. Wu's experiment focused on the underlying symmetry. If parity was not conserved, then there ought to be weak interactions that took place in a way that would not look the same in a mirror. Basically, Wu and the scientists from the National Bureau of Standards looked for asymmetry in the beta decay of a radioactive element, cobalt 60. First they had to line up all the atoms so that each one was spinning in the same way and had the same pole pointing in a given direction. To do this, they first cooled the sample of cobalt 60 to a very low temperature, near absolute zero. This reduced the movement of the atoms. A magnetic field was used to line up the atoms according to their spins. The magnetic field could not line up all the atoms, but it could insure that more than half the atoms had their north poles pointing in the same direction. In beta decay, a weak interaction, a neutron emits an electron and an antineutrino as it decays to a proton. It was known that the electron would be emitted in the direction of one of the poles of the nucleus of the atom. If mirror symmetry were invariant, an electron would have an equal chance of being emitted from the north pole as the south pole. Wu and her colleagues found, however, that 40 percent more electrons were coming from the south poles. The effect was strong enough to convince the world that mirror symmetry was not invariant in beta decay and, by implication, that parity was not conserved in weak interactions.

Leon Lederman was at the Chinese Lunch that day, and it soon occurred to him that a variation of some experimental work he was doing at the time should also detect nonconservation of parity. Starting that Friday night, he and R. L. Garwin worked through the weekend and, by Tuesday morning, Lederman was able to telephone Lee with the news that he had also shown that parity was not conserved in weak interactions, or, as Lederman put it "parity is dead." At 2 A.M. on Wednesday, Wu's group also concluded their experiment, celebrating the fall of parity with champagne in paper cups. At the same time, physicists at the University of Chicago were obtaining results that looked like parity nonconservation.

Word of these results swiftly spread throughout the physics community. There was enough curiosity about what was going on that the Columbia physics department felt that a formal announcement was needed, so they held a press conference on January 15, 1957 to announce the fall of parity to the world, duly reported the following day in the *New York Times* and other papers. There was considerable confusion, however, as few if any of the reporters had the slightest notion of what parity was or what its fall might imply. Since the physicists were making such a big thing out of it, the reporters tended to think that knowing that parity was not conserved would unlock all the riddles of physics.

Well, it was a big thing, but not that big. Physicists all over the world were astonished by the fall of parity. Feynman lost his 50 dollars. I. I. Rabi noted that "In a certain sense, a rather complete theoretical structure has been shattered at the base, and we are not sure how the pieces will be put together." Pauli, working in Zurich, had expressed his views on January 17, 1957: "I do *not* believe that the Lord is a weak left-hander," but 10 days later, after he had read the scientist's reports of the actual experiments, he had to admit that he was wrong. Pauli wrote, "I am shocked not so much by the fact that the Lord prefers the left hand as by the fact that He still appears to be left-right symmetric when He expresses Himself strongly. In short, the actual problem now seems to be the question: Why are the strong interactions right-and-left symmetric?"

By the end of 1957, Yang and Lee had won the Nobel Prize in physics.

Parity in an Antigalaxy

How could we use nonconservation of parity to tell our alien what we mean by right and left? One way to do it would be to repeat the Wu experiment. From the Wu experiment, it is possible to tell the north pole from the south pole of the nucleus of an atom of cobalt 60. Working backward, the Wu experiment also indicates which is the north and the south pole of an electromagnet, since the magnetic field determines the orientation of the nuclei. If you know which is the north pole of a magnet, then you can use a right-hand rule or a left-hand rule (depending on whether you assume current goes from positive to negative or vice-versa) to tell which way the magnetic field of a current will cause a compass to point. (Remember that the south pole of a magnet points toward the north pole of the Earth, so what we commonly think of as a compass has an arrowhead on the south pole.) But you don't have to have a right hand or a left hand if you have an actual compass. We assume that if the alien can build electromagnets and can achieve temperatures near absolute zero, it also has learned about electrons, so it would be able to identify the negative and positive poles for the electric current. Thus, having learned north from south as a result of repeating the Wu experiment, all that would be needed would be to place a compass above a wire carrying a current from a negative pole near you to a positive pole farther away. The compass would point to the left.

That would seem to solve the problem, but there is a further complication. As noted in Chapter 5, there is at least a possibility that galaxies of antimatter exist. Suppose that our alien is actually an antialien on one of those galaxies. In that case, what we describe to the alien as electrons would be, unknowingly to either side, antielectrons—positrons. Then what would happen?

Anticobalt 60 would release positrons in decay instead of electrons. Experiments with the decay of another isotope of cobalt, cobalt 58, which decays by releasing positrons, show that they preferentially emerge from the north pole, not the south. Thus, the Wu experiment if conducted with anticobalt would be expected to produce the reverse result of what was intended; what we have told the alien is the north pole of the magnet is actually the south. A compass built according to the instructions just given would point south in an antigalaxy.

Furthermore, an anticurrent would flow from the positive to the negative poles in the wire, but the antialien, not knowing that it lives in an antigalaxy, would label the pole from which the current flows as the negative pole. We tell the antialien to arrange the current so that the particles that he thinks are electrons, but which are really positrons, are flowing away from it. In electromagnetic theory, however, a flow of positrons in one direction has the same effect as a flow of electrons in the other direction. The electromagnetic field produced by the flow would be oriented in the same way as the field would be if there were electrons flowing from what the alien has incorrectly labeled the positive pole to the incorrectly labeled negative pole. Thus, the negative flow of current is actually toward the antialien instead of away from it. A correctly labeled compass placed above the wire would point to the right. Therefore, when the incorrectly labeled compass is placed over the wire, it will point to the left.

This may be a little difficult to picture. Try it out with your hands. Point your left-hand thumb away from your body with your palm vertical and fingers pointing up. Then curl your fingers. They point to the left, indicating that a current of electrons flowing away from your body would cause the north pole of a compass placed above the current to point to the left. Now reverse the direction of the current by pointing your left-hand thumb at your body with your fingers pointing down. When they are curled, your fingers still curl to the left, but now they are below the current. If they were above the current, they would point to the right—with a flexible-enough wrist, you can twist your left hand around to observe this. Thus, the north pole of the magnet would point to the right, but the south pole—incorrectly labeled as north—would point left. Reversing both the direction of the negative current and the poles of the magnet gives the antialien the same understanding of left and right as it does the alien.

Thus, it is possible to communicate top and bottom, near and far, . . . and right and left—even to an antialien.

Another thought experiment, however, would seem to contradict this. It can be shown that a disk of aluminum coated on the top with cobalt 60 that is suspended from a string through the center of the disk and in a strong magnetic field will have a slight, unmeasurably slight but still there, tendency to rotate counterclockwise. Therefore, a similar disk coated with anticobalt 60 would tend to rotate clockwise. Since the counterclockwise-clockwise description is equivalent to the left-right distinction, it would appear that an antialien should reverse left and right. What is wrong?

In repeating the Wu experiment, the antialien made *two* reversals—the poles of the magnet and the charge of the current. In the disk experiment, however, there is only one reversal, the reversal obtained by using anticobalt instead of cobalt.

Although the force that tends to cause an aluminum disk to rotate is exceedingly small, it theoretically can be measured, and physicists have shown themselves to be quite ingenious in measuring very small forces or effects. Thus, the nonconservation of parity in weak interactions allows us to communicate two pieces of information to an alien that we could not communicate if parity were always conserved; the difference between right and left and the difference between matter and antimatter. If the Wu experiment and the disk experiment give different results, the alien lives in a pocket of antimatter, and there is not much point to trying to arrange a live meeting. If the alien came to Earth, or if we went off to visit its galaxy, the result would only be a giant explosion.

Thoughts along these lines soon led Yang and other theoreticians to propose a way to "save" the lost conservation law. Since parity is not conserved in weak interactions, it is possible to tell whether or not you are looking at a particular interaction or its mirror image. Suppose, however, you had a very unusual mirror, one that not only reversed left and right, but also changed every particle into its antiparticle. Although such a mirror is impossible to construct physically, it is easy to construct mathematically through the appropriate transformations. The mathematical transformation that changes all particles into their antiparticles is called *charge conjugation.*

Shortly after the fall of parity, it seemed sensible to check whether or not charge conjugation behaved symmetrically. Before parity's fall, it seemed reasonable to assume that if you replaced each particle in an interaction with its antiparticle, the new interaction would look just like the first. Experiments conducted at the University of Liverpool later in 1957 showed that this was not the case.

To save some form of symmetry, what the theoretical physicists next proposed is that the combination of mirror inversion and charge conjugation would be conserved in all particle interactions, weak or otherwise. This came to be known as CP conservation (for Charge conjugation-Parity). CP conservation is aesthetically satisfying because it means that nature does not make arbitrary distinctions.

It is even possible to think of a model for CP conservation that would explain why it happens, although there is little evidence that this model really describes reality (just as the Bohr model of the atom in which electrons travel in orbits and jump from one orbit to another does not describe the reality of the atom, but still provides a useful mental picture). Suppose that charge is yet another kind of spin, with a negative charge and a positive charge being simply spins in the opposite direction. If you could see this charge-spin in a mirror, the reflection would have the opposite charge-spin. Now also suppose that charge-spin has something about it that influences the direction of weak interactions. To get a little closer to reality, you need a form of charge-spin that carries no charge, so that a neutron and an antineutron can also be distinguished by charge-spin, even though both are neutral. The nucleus of cobalt 60 would have a positive charge-spin and emit most electrons from its south pole, but its mirror image would show a charge-spin that was negative and would emit most electrons from its north pole. This would make sense if the charge-spin had some influence over the direction in which weak interactions happen.

There is a more physical way to look at CP conservation. All weak interactions involve either neutrinos or antineutrinos or both. It is also believed that a neutrino or an antineutrino is almost all spin, this time spin of a more conventional type. Since these form a neutral particle–antiparticle pair, their chief difference seems to be the direction in which they spin. After charge conjugation, the neutrinos are replaced by antineutrinos and vice versa. In a ordinary mirror image of a weak interaction such as beta decay, the antineutrino carrying spin of one type away from the original reaction is replaced with a neutrino carrying spin of the reverse type. Thus, in an ordinary mirror, parity is not conserved. But in a mirror that also performs charge conjugation, the reflection of beta decay would show an antineutron emitting a positron and a neutrino. If the neutrino has its apparent spin reversed by the mirror, the charge-conjugating mirror image will look exactly like the original beta decay.

Seven years after the fall of parity, however, physicists were able to conduct specific experiments to determine whether or not CP conservation was really a feature of reality, or just another aesthetic notion, as parity conservation had turned out to be. James W. Cronin, Val L. Fitch, and their colleagues at Princeton University re-examined the villains in the tau–theta paradox. By this time, it was common to call the former *tau* by the name of *kappa two zero,* or K_2^0, while *theta* had been rechristened *kappa one zero,* or K_1^0. Today the same symbols are used, but read "Kaon two zero" and "Kaon one zero." Actually, it is a little more complicated than that, since *tau* and *theta* were also used to refer to charged states of the kappa meson, but the essence of tau-ness and theta-ness is carried by the two forms of the neutral K meson, since K_1^0 always decays to two pions, while K_2^0 decays to three pions according to the much-better-understood particle physics of the 1960s.

What Cronin and Fitch found was another surprise, for in those 1964 experiments at Princeton University, CP was not conserved in certain weak interactions. In a charge-conjugating mirror (mathematically, that is), the mirror-image interaction should occur with equal probability as the original interaction according to CP conservation, but that was not always the case.

The implications of this discovery are more surprising than the discovery itself.

Reversing Time's Arrow

Unlike the Wu experiment, which connects with parity in a very direct way, the Cronin-Fitch experiment revealed a very subtle disturbance in the symmetry of the universe. It had been shown that the neutral K, or K^0, and its antiparticle fluctuate back and forth; that is, a K^0 can decay into two pions that can merge to form the anti-K^0, which can also decay into two pions that can merge to form the original K^0. Each of these two states then becomes mixed, and the mixed state behaves somewhat as a single particle. It probably helps to think of the particles as waves to grasp how this might happen, since it is much easier to picture waves as mixing to form another wave then it is to picture particles mixing to form another particle. The merger can happen in two different ways. In one of them, the two wave functions add to produce the wave-particle recognized as K_1^0. In the other, the wave functions subtract to form K_2^0. Neither of these particles actually exists—they are just the superimposition of the neutral K and its antiparticle. The result of this mixing, however, is that when a physical process produces K mesons, producing as is expected the same amount of Ks and anti-Ks, half of the decays will be by the K_1^0 path and the other half will be by the K_2^0 path. From this interpretation of how the K and anti-K behave, it can be shown that CP conservation will dictate that the K_1^0 path is to decay into two pions, but the K_2^0 state can never decay according to this path. Similarly, CP conservation implies that K_2^0 can decay into three pions, but K_1^0 cannot.

These two decay processes take different amounts of time. It takes longer to decay into three particles than into two; specifically, it takes K_2^0 about 200 times as long as K_1^0, although both decay into pions in less than a billionth of a second. Because particles are moving so fast when they leave their images in particle detectors, the distance they have traveled before the decay occurs can easily be measured. Particle detectors do not see most neutral particles easily, so what appears for the neutral kaons consists only of charged decay products. To complicate the picture a little more, K_2^0 decays into three neutral pions, so you don't see those either. Finally, the neutral pion decays into two gamma rays (photons), which fortunately can be detected.

The energy of the decay products tells you the mass of the unseen neutral particle, by the law of conservation of mass-energy. Both K and anti-K must have the same mass, since one is the antiparticle of the other. But K_1^0 and K_2^0 have different masses, since one is an additive combination and the other a subtractive combination. Therefore, it is possible to tell one from another.

We now are in a position to understand the Cronin-Fitch experiment. They produced a beam of kaons. As always, there were the same number of Ks and anti-Ks. The beam was directed into a particle detector. At a certain distance, because the K_1^0 has a faster decay mode, the beam contained K_2^0 states only. Therefore, beyond that distance, it would be expected that all the particles would decay into 3 pions. But they found that about 0.1% of the particles decayed into 2 pions. These could not be unusually long-lived K_1^0 states because the mass corresponded to the K_2^0 state.

The theory based on CP conservation did not accord with the results. Therefore, the efforts to save symmetry in physical theory had not been sufficient. There is another way out, however. A third symmetry that is in many ways like parity and charge conservation is symmetry in time. Specifically, physicists had reason to believe that if you observed a possible interaction in a "mirror" that reversed time (instead of left and right as ordinary mirrors do or instead of particle and antiparticle as charge-conjugation mirrors do), the mirror-image interaction would also be physically possible.

A good example is the production of an electron and a positron from an energetic photon (gamma ray). In the time-reversing mirror, the image would be an electron falling into the positron hole, producing the photon. Both interactions are observed in real life. Recall that it was this pair of interactions that suggested to Feynman that a positron could be thought of as an electron moving backward in time.

In theory, all physical processes would also have this property of symmetry in time. Thus, time symmetry has also been one of the bases of physics.

Around 1950, furthermore, a group of theorists had proved from very basic principles, from the Dirac equation that Dirac had noted was more intelligent than he was, that the combination of parity, charge conservation, and time had to be conserved. Since this result, known as the CPT theorem, was a direct result of the basic equation of quantum electrodynamics combined with the special theory of relativity, physics would be in very bad trouble if it were not true. Therefore, the experimental results of the Cronin-Fitch experiment should not be interpreted as violating the CPT theorem.

The results did show that a small percentage of events violated CP conservation, however. If you reflected these events in an ordinary mirror and in a matter-antimatter-reversing mirror, the image event would be impossible. But if the image was reflected in a time-reversing mirror, the CPT theorem guarantees that the final image event would be possible. Therefore, once in a while an event occurs that could not happen if time were going backwards.

Physicists think this is very important, although they have yet to work out any implications beyond the experiment with neutral K mesons.

One thought experiment may help clarify this experiment. Think again of communicating with our distant alien who lives in a galaxy far, far away. We now know how to tell the alien which is right and which is left. We can also use that information to tell the alien whether or not the galaxy it lives in is made of matter or antimatter. Before 1964, however, we would have no way to tell whether or not time in the alien's galaxy flowed the same way that it does here. (It has been argued that communication would be impossible between people living forward in time, as we think we do, and people living backward in time.) Now we know that the alien could conduct the Cronin-Fitch experiment. If it got the same results as we do, then time has the same direction in the alien galaxy as it does here.

8

The Group's All Here

One cannot study any physical system for very long before finding regularities or symmetries and, even though the system may be complex, one expects that the regularities will have a simple explanation. This basic optimism, which pervades not only physics, but science in general, is justified in the case of symmetries because there is a theory of symmetry which has application in almost all branches of physics and especially in quantum mechanics.

J. P. Elliot and
P. G. Dawber

"A likely story indeed!" said the Pigeon. "I've seen a good many little girls in my time, but never one with such a neck as that! No, no! You're a serpent; and there's no use denying it. I suppose you'll be telling me next that you never tasted an egg!"

"I have tasted eggs, certainly," said Alice, who was a very truthful child; "but little girls eat eggs quite as much as serpents, you know."

"I don't believe it," said the Pigeon; "but if they do, then they're a kind of serpent: that's all I can say."

This was such a new idea to Alice, that she was quite silent for a minute or two. . .

Lewis Carroll

You are now ready to complete your understanding of Klein's definition of a geometry as the set of properties that are invariant under a particular *group* of transformations, which was introduced in Chapter 3. The notion of a *group* is so simple that anyone can understand the basic idea, but it is so deep that mathematicians have been working out its puzzles for the last 150 years. One of the major milestones of group theory was reached as recently as in 1980. Furthermore, knowing what a group is becomes essential to understanding the conclusion (at least for now) of the story that was begun in Chapter 6, the story of how symmetry has come to provide our current understanding of particle physics.

The plan is to work through the mathematical idea of a group, and then to see how physicists applied this idea, leading to a new theory of the weak force and then to a new theory of the strong force that together became the *standard model* for particle physics, and finally to new and unproved theories that *almost* explain the universe from beginning to end in terms of symmetry.

Let Me Count the Ways

Today, group theory is usually developed in popular works of mathematics with illustrations from geometry and in more serious works as part of abstract algebra, but its earliest roots were in counting. Counting theory, or combinatrics, is an interesting branch of mathematics that is usually studied along with probability theory because of its many applications in that area. Strictly speaking, however, counting theory is an isolated island in mathematics. Counting theory asks the question "How many?" about various classes of objects. For example, in trying to find the cure for a rare disease, a researcher decides to test a different treatment on each of three patients. How many ways can the treatments be allotted to the patients?

The mathematician Augustin-Louis Cauchy (1789–1857) investigated one particular aspect of counting theory thoroughly in the 1840s. He was concerned with problems similar to the example given previously concerning the treatment for a rare disease. Call the three patients A, B, and C. For the first treatment, any one of the three patients may be assigned. Once that has been done, there are only two choices of treatments for the second patient; and once that assignment has been made, the last patient will get the leftover treatment. Thus, there are six different assignments that can be made. We can conveniently show these assignments by using just the letters for the patients, letting the order of the letters designate the treatments. In that case, the assignments turn out to be ABC, BAC, CAB, ACB, BCA, CBA.

Cauchy looked at this problem from a different angle—always a good technique for mathematical discovery. He considered how to get from one assignment to another. For example, to get from ABC to BAC you interchange A and B. Such an interchange is called a *permutation.*

There is a standard method used to indicate permutations. The objects to be interchanged are shown in parentheses, with commas between them. The notation used to indicate the interchange of A and B is (A, B). The same change could also be shown as (B, A).

When (A, B) is applied to CAB the result is CBA. When (A, B) is applied to ACB, the result is BCA. None of the results of applying (A, B) can be other than the six assignments already given. It is easy to see that permutations of A, B, and C will always result in one of the six original arrangements and not some totally different kind of assignment altogether, because we know that those six assignments are the only ones possible. However, a little experimentation will show that starting from ABC, the permutation (A, B) applied once will give BAC; and starting from BAC, the permutation (A, B) gives ABC again. It does not reach any of the other assignments.

Now look at the permutations that give the other assignments. Besides (A, B), there are two others that interchange two of the letters, (A, C) and (B, C). Will these three produce all six of the assignments from ABC? Yes, but some have to be used more than once since some of the assignments involve double interchanges. To get from ABC to BCA you need to interchange A and C (giving CBA) and then interchange B and C. While getting from ABC to BCA can be accomplished by two interchanges, it is more convenient to have a single operation that produces BCA from ABC. The standard notation for this kind of permutation is (A, B, C). This notation is a little confusing.

Remember when you were looking at the 15-puzzle in Chapter 4, there was also the small version, the 3-puzzle. Think about the way that the numbers 1, 2, and 3 chased each other (and the hole) around the four corners of their small world in the 3-puzzle. Although there were 4 different positions possible, the 1, 2, and 3 could not interchange with each other or with the hole. The notation (A, B, C) is based on the same idea. It represents an interchange that leaves A, B, and C in the same order, although C is now first. That is, A is changed to B, B is changed to C, and C is changed to A.

There are two other assignments that still cannot be reached by a single permutation, once you have introduced (A, B, C). One of them is CAB. It can be obtained with (A, C, B) which means to replace A with C, C with B, and B with A. The other one is ABC itself. In this case, every letter is kept identically the same, so the standard notation is I (for identity). This completes the list of permutations that interchange ABC into all of its forms.

I, (A,B), (B, C), (A, C), (A, B, C), (A, C, B)

Once Cauchy had gotten this far, he threw away the assignments and worked with just the permutations. It has already been noted that the permutation (A, B, C) is the same as (A, C) followed by (B, C). This can be written as (B, C) * (A, C) = (A, B, C). (You may want to read the operation * as "preceded by"; the convention is to write the operation that is performed first on the right.) Similarly, you can check to see that (A, C) * (B, C) = (A, C, B). This is a surprise! Combining permutations is not commutative, since (B, C) * (A, C) is not equal to (A, C) * (B, C). A lot of the familiar operations, such as addition and multiplication, are commutative ($a + b = b + a;\ a \times b = b \times a$), although not all of them are. For example, division is not commutative; in general, a divided by b does not equal b divided by a. Intuitively, however, you would expect * to be commutative, but it is not.

It is helpful to organize the results of combining two permutations into a table, similar to the addition or multiplication table. In this table, we will simplify the reading of the table by assigning a number to each of the permutations. If (for reasons to be made clear soon) we assign them as follows:

0 = I
1 = (A, B, C)
2 = (A, C, B)
3 = (B, C)
4 = (A, C)
5 = (A, B)

the table can be shown as follows:

*	0	1	2	3	4	5
0	0	1	2	3	4	5
1	1	2	0	4	5	3
2	2	0	1	5	3	4
3	3	5	4	0	2	1
4	4	3	5	1	0	2
5	5	4	3	2	1	0

To read a particular combination from the table, you start with the heading for a column and look down until you find the row. Thus $1 * 3 = 5$, while $3 * 1 = 4$. So, although the table looks like an addition table, it is not.

If you look at another familiar rule for addition and multiplication, however, you find that it is true for $*$. If a, b, and c are permutations, it is always true that $a * (b * c) = (a * b) * c$. Check it out on the table: For example, $3 * (4 * 5) = 4$ and $(3 * 4) * 5 = 4$. This property is known as *associativity*.

The Legend of Group Theory

Here is the legend. Around dawn on May 30, 1832, a 20-year-old Frenchman was killed in a duel. Although this young man was an immensely talented mathematician, he had been rejected by the powers-that-be of French education and mathematics because of his radical political views. On the night before he died, he took the fundamental ideas of Cauchy's theory of permutations, expanded them, and created a new branch of mathematics, called *group theory*. Group theory has since become the dominant mathematical idea and the guiding force in physics of our time. His presentation of the theory was necessarily sketchy, since he had little time in which to explain his new theory before going forth to meet his challenger. It has taken 150 years for mathematicians to work out the implications of that one paper.

Much of this legend is true, although heavily romanticized. Evariste Galois, the young Frenchman, was among the creators of group theory, but he did not do it that night, and his work was not rejected because of politics. Furthermore, it appears clear that group theory would have been created by someone around that time in any case. It was in the air. Mathematicians would have been working out the intricacies of group theory for the past 150 years if Galois had never lived—and certainly not all of the mathematics was implied in the papers found on the morning of May 30, 1832. On the other hand, the achievement of Galois is quite remarkable. For one thing, he developed group theory in the process of providing a general solution to a problem that had baffled mathematicians for 300 years. It is not nearly so clear that any other mathematician would have solved this particular problem with these methods.

What is the basic idea of group theory and how important is it anyway? For the latter question, one can quote from a well-known textbook: "If a person is not ready and anxious to explore the concept of a group further, we recommend that he reconsider his relationship to mathematics." This conclusion of H. P. Griffiths and P. J. Hilton in *A Comprehensive Textbook of Classical Mathematics* may be slightly prejudiced in favor of group theory, but it is certainly true that group theory crops up in most of mathematics. The reason that group theory can appear anywhere is that group theory is so simple in essence, although complex in practice. Simple ideas are the basis of most of mathematics (except perhaps for calculus).

The first notion to be grasped is the same as the one that Cauchy found held true for his permutations. If you combine two things in a group by the rule of combination you are using, the result will be a thing in the group. This fundamental property is called the *group property,* although it is also known as *closure.*

It is easy to find sets of things that have the group property. Numbers are an obvious example. If you make addition your rule of combination, for example, and consider the set of whole numbers (0, 1, 2, 3, . . .), combining two members of the set by the rule will always give you another member of the set. The sum of any two whole numbers is a whole number, so the set of whole numbers has the group property for addition. The same is true for whole numbers when the rule is multiplication, but it is not true when the rule is subtraction, since, say, $3 - 5$ is not a whole number. Thus, the whole numbers have the group property for addition and multiplication, but not for subtraction (or division either).

Transformations acting upon geometric objects can have the group property also. One common example is an equilateral triangle that is rotated counterclockwise about its center by 120° (Figure 8–1). Each rotation brings the triangle into coincidence with itself, with only the letters at each of the vertices changed.

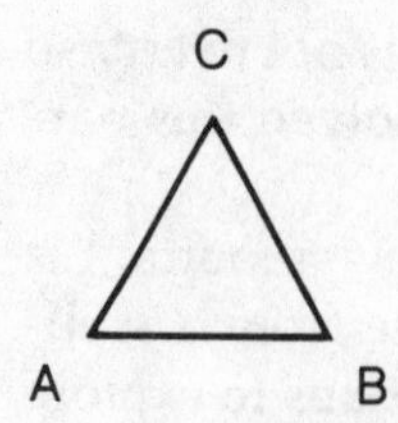

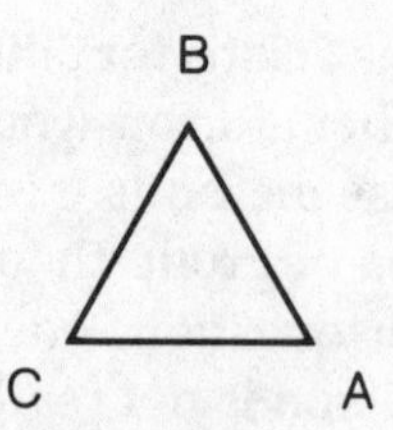

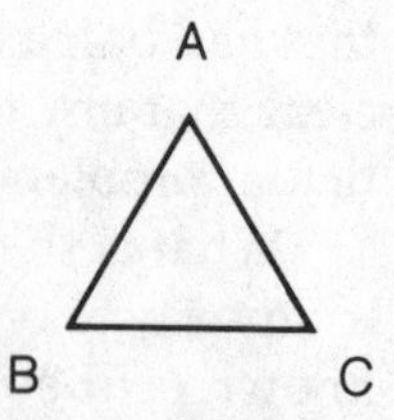

FIGURE 8–1
Rotating an equilateral triangle a third turn results in the same triangle with different letters at the vertices.

The group property is also preserved if you permit another kind of transformation of the triangle as well, reflection about an altitude. Since the mirror images of the three positions shown in Figure 8–1 are all different from the original triangle (as can be seen by the labeling of the vertices), the result is three more triangles that also coincide with the original one. Therefore, the triangle with these two operations (counted as a single rule) has the group property (Figure 8–2).

Notice that the different positions of the triangle can all be indicated simply by the letters of the three vertices. Thus, if you

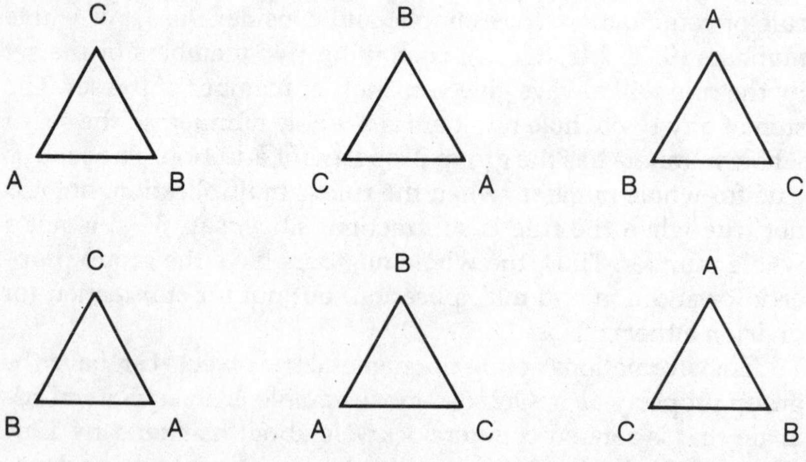

FIGURE 8–2
A combination of rotation and reflection produces six different ways of labeling the vertices of the triangle.

start reading in the lower left-hand corner and proceed counter-clockwise, they are

ABC, CAB, BCA, BAC, ACB, CBA

Although the order is slightly different, you should recognize this set as the same one you saw when considering the case of trying drugs on patients A, B, and C or when following Cauchy's train of thought.

ABC, BAC, CAB, ACB, BCA, CBA

Thus, another set with the group property is the set of permutations of three objects. We have returned to where Cauchy started. You will recall that Cauchy found that the permutation operation was not commutative. Thus, whatever a group has besides the group property, it is not commutativity.

Three Other Group Properties

Aside from the group property, there are three other properties that all groups (by definition) have in common. Just as Cauchy found it convenient to include the permutation that leaves everything in its place, all groups must have an *identity element,* a member of the group that when combined with another member according to the group rule leaves the group unchanged.

Looking back to the sets of numbers that had the group property, you can note that 0 is the identity element for the whole numbers with the addition rule and 1 is the identity element for the whole numbers with the multiplication rule.

With this in mind, think about the first triangle example (Figure 8–1 with just rotations) in a different way. The members of the group make more sense as the transformations, not as the triangle itself. This is very similar to the shift Cauchy made when he turned from looking at arrangements to looking at the permutations that made the arrangements. The triangle group of rotations then becomes

T_0 is rotating the triangle 0°
T_1 is rotating the triangle 120°
T_2 is rotating the triangle 240°

In this case, the rotation of 0°, or T_0, is the identity element. The rotations have the group property if the operation is identified as "preceded by," which will be shown as *. Notice that $T_1 * T_2 = T_0$ and $T_2 * T_1 = T_0$. Since T_0 does not change either T_1 or T_2 no matter whether it precedes or follows, this particular set and operation form a commutative group. This is acceptable. Groups can be either noncommutative or commutative.

When you add to this set the three transformations that are reflections about the three altitudes of the triangle, you still have one identity element, T_0, but you have a set of six transformations with the operation "preceded by," or *. (See Figure 8–2.) Call the three new transformations T_3 (reflection about the altitude from C), T_4 (reflection about the altitude from A), and T_5 (reflection about the altitude from B). Conventionally, in dealing with these transformations, the three axes remain unchanged during these rotations. Thus, $T_4 * T_1$ means to rotate the triangle counterclockwise by 60° and then to reflect about the altitude that starts at the lower left-hand vertex of the triangle. Therefore, in the operation $T_4 * T_1$, the reflection is no longer about the altitude through A, but instead is about the altitude through C, since T_1 has moved C into the lower left-hand corner of the triangle.

Remember that in the *Erlanger Programm,* Klein defined a geometry as the set of properties that are invariant under a particular *group of transformations.* We are getting very close to the meaning of Klein's definition.

The third property of a group is that for any member of the group there must be another member such that when these two specific members of the group are combined, the result will be the identity element of the group. This property seems rather more abstract than the other two, and it effectively knocks out many candidates for groupship. It does not eliminate the group of triangle rotations, however. Let us call that group $\mathbf{G}_1$. $\mathbf{G}_1$ can be described as $\{T_0, T_1, T_2, *\}$; that is, by listing the elements of the group in curly braces along with the group operation. To show that G_1 has the third group property, it is only necessary to observe that

$$T_0 * T_0 = T_0$$
$$T_1 * T_2 = T_0$$
$$T_2 * T_1 = T_0$$

Mathematicians call T_2 the *inverse* of T_1, since their combination by the group operation is the identity element for the group. Every member of a group must have an inverse. The inverse of a group member is symbolized by the exponent -1, so another way to show this relationship is

$$T_1^{-1} = T_2$$

Likewise, T_1 is the inverse of T_2, so $T_2^{-1} = T_1$. The transformation T_0 is its own inverse, or $T_0^{-1} = T_0$. (The identity element must always be its own inverse for any group.)

Now include T_3, T_4, and T_5 to G_1 to make the group $\{T_0, T_1, T_2, T_3, T_4, T_5, *\}$, which can be called G_2. Can you find the inverses of the new group elements in G_2? This turns out to be easy, since each of these transformations is its own inverse.

$$T_3 * T_3 = T_0, \text{ so } T_3^{-1} = T_3$$
$$T_4 * T_4 = T_0, \text{ so } T_4^{-1} = T_4$$
$$T_5 * T_5 = T_0, \text{ so } T_5^{-1} = T_5$$

One way to show a finite group such as G_2 is to show a table similar to a multiplication table for numbers 1 through 10.

*	T_0	T_1	T_2	T_3	T_4	T_5
T_0	T_0	T_1	T_2	T_3	T_4	T_5
T_1	T_1	T_2	T_0	T_4	T_5	T_3
T_2	T_2	T_0	T_1	T_5	T_3	T_4
T_3	T_3	T_5	T_4	T_0	T_2	T_1
T_4	T_4	T_3	T_5	T_1	T_0	T_2
T_5	T_5	T_4	T_3	T_2	T_1	T_0

In this table, the operation * is read as "column preceded by row." For example, $T_3 * T_4$ is found in the column headed T_3 and the row headed T_4, so it is T_1; while $T_4 * T_3$ is in the T_4 column and the T_3 row and is T_2.

Finally, the last required property of a group is that it be *associative;* for example, $T_1 * (T_2 * T_3) = (T_1 * T_2) * T_3$ must be true. For a finite group, it is possible to check all possible combinations of three elements of the group to determine whether or not the associative law is maintained. Often the group operation is known from other contexts to be associative. Both of the groups based on rotations or rotations and flips of an equilateral triangle are associative. In the case of rotations alone, associativity emerges from the operation itself. In the larger group of rotations and flips, you may want to check some of the possible combinations to convince yourself that this group really is associative. If it were not, it would not be a group.

One Thing Looks Just Like Another (Again)

A close study of the table for $\mathbf{G}_2$ will show the relationship between this group and the permutations of A, B, and C that Cauchy studied. In fact, the tables for the operations are almost identical. The only difference is that these are transformations indicated by T with a subscript, while the permutations are shown as numbers, which turn out to be the same as the subscripts in the table for $\mathbf{G}_2$. For all practical purposes, you do not need two separate tables of operations for these groups, since either table tells you the various combinations that make up the group.

At the same time, permutations are not exactly the same thing as transformations, so the two tables actually do describe two different groups. Mathematicians say that the two groups have the same shape, using the term *isomorphic.* A rule that connects one group to another in a one-to-one matching between elements is an *isomorphism.*

The group formed by the permutations of three objects is known to mathematicians as $S(3)$. In fact, the name $S(3)$ may be used for any group that is isomorphic to the group of three permutations, so the group $\mathbf{G}_2$ is also an $S(3)$ group. Similarly, the group of permutations of four objects would be $S(4)$.

When the Pigeon told Alice that she must be a serpent because both Alice and serpents eat eggs, the Pigeon was forming a kind of isomorphism between two very different sets. Similarly, when two groups are isomorphic, you know that if something is true about one of them, then the same thing will be true about the other. For example, since the group of permutations is not commutative, you know that the group of transformations will also not be commutative. For another example, you know that the three rotations of the triangle formed a group by itself, $\mathbf{G}_1$. Therefore, the three permutations of the letters A, B, C that correspond to the rotations must also form a group by themselves. These are I, (A, B, C), and (A, C, B), which represent the permutations that keep the letters A, B, and C in the same order. Such a group within a group is called, logically enough, a *subgroup*. With what may seem like less logic, except to mathematicians, the whole group is also considered to be a subgroup of itself. In practice, this works out very well, although you need first to rid yourself of the idea that *sub-* always means "contained within" or "smaller than." In fact, in mathematics, this is almost never the case. While some subsets, for example, may be smaller than the set of which they are subsets, the set itself is also a subset of itself. The same situation holds for groups and subgroups.

Not every group with six elements is isomorphic to the group of permutations. Consider, for example, the group that consists of the six rotations of a hexagon that keep it in place (each rotation is 60° counterclockwise more than the preceding one). If we call these rotations, in order, 0, 1, 2, 3, 4, and 5, the table looks like this:

*	0	1	2	3	4	5
0	0	1	2	3	4	5
1	1	2	3	4	5	0
2	2	3	4	5	0	1
3	3	4	5	0	1	2
4	4	5	0	1	2	3
5	5	0	1	2	3	4

No amount of reassigning the names of the rotations in this table to other numbers will make the table the same as that for the transformations of the triangle or the permutations of three letters. This is a different group altogether. For one thing, it is a commutative group.

Particles Come in Groups

When we left particle physics, it was still somewhat in disarray. There were too many particles and too many forces. In addition to the strange particles that had been first seen in 1946, physicists were using more and more powerful particle accelerators to produce all kinds of new particles, a plethora that came to be known as "the particle zoo." Indeed, Enrico Fermi remarked around this time that if he had wanted to be in a field that required knowing the characteristics of so many different things, he would have become a botanist. Furthermore, although there was considerable understanding of the four fundamental forces (strong, electromagnetic, weak, and gravitational), it seemed that they ought to be connected with each other in some way. One force with different manifestations ought to be sufficient to hold the universe together; did God really need four different ones?

Let's go back to where we left off in Chapter 7, at the end of the 1950s and the start of the 1960s. Strange particles are still the focus of attention of most of the theoretical particle physicists. Furthermore, the experimentalists now have their new toy; the giant particle accelerator. Although the first particle accelerators to be successful had been built in 1932 by Ernest O. Lawrence in the United States and by James Cockcroft and Ernest Walton in England, they did not achieve the energies necessary for the creation of previously unknown particles.

Since mass and energy are connected by $E = mc^2$, it follows that $m = E/c^2$. Since c is a very large number, E must be also relatively large to produce even a small m. Consequently, it is common to speak of the very small mass of a subatomic particle in terms of the energy needed to generate it. For ordinary objects we think in terms of grams, a measure of mass, but for very small objects we think in terms of a measure of energy.

The measure of energy chosen is a very small one. Particle mass is measured in electron volts and related units. The energy of a single electron volt is the energy acquired by one electron when it is accelerated by one volt. This is not very much energy. Expressed in the terms used for the physics of the macroworld, it is 0.0000000000000000001602 Joule, or 0.00000000000000000011816352 foot-pounds. The electron volt is actually too small even for subatomic particles. Multiplying either of these amounts by a million, you get 1 million electron volts, abbreviated 1 MeV. One MeV is a convenient size to measure the masses of subatomic particles. A proton, for example, has a mass of 938 MeV.

To get some kind of common perspective on this, remember that a foot-pound of energy is used to lift a weight of one pound vertically a distance of one foot. But there are an enormous number of protons and neutrons in an object that weighs one pound, and each one has a mass that is equivalent to about 0.000000000001 of that energy.

By the late 1950s, it was possible to produce energies that were from three to six times as great as needed to produce a proton out of pure energy. Before that time, the most massive particles known were the lambda particles found in cosmic rays—and they were only a little less than 20 percent larger than protons. As physicists pushed the energies up, they did not just get heavier versions of the known particles. Instead, they got definite new particles at specific energies. Although some of these had been known from work with cosmic rays, it was the first time that physicists could obtain an abundance of the particles to study. There was one they called the positive sigma, for example, that weighed in at 1189.4 MeV. Nearby was a negative sigma at 1195.3 MeV, and in between a neutral sigma at 1192.5. A little further up the scale were the 1315-MeV neutral xi and the 1321-MeV negative xi. The first unexpected particle to appear in a particle accelerator was the delta, which showed up in

1952 with a mass of about 1236 MeV. Each of these particles had definite characteristics, a definite spin for example, or a definite lifetime before decay into definite sets of smaller particles.

As the experimentalists produced this particle zoo, it became the problem of the theoreticians to explain why these new particles occurred with the particular properties they had and not some other set of properties.

A way to handle this problem had existed for over 60 years, but not many physicists knew about it. In 1894, for his doctoral thesis, a French mathematician, Elie Cartan, had classified all known finite groups—an effort that was so prodigious that it caused most mathematicians to lose interest in finite groups as a field of investigation. Not that they did not use finite groups in various ways, but for a period of time it was just not very interesting to work on something in which most of the problems had been solved. In the 1930s, however, interest revived somewhat as various mathematicians extended Cartan's work, studying the 18 families of simple groups and 26 groups that oddly enough did not belong to any of the families.

This effort came to a satisfactory conclusion rather unlike that found elsewhere in mathematics or science; in 1980, the last of the possible groups, one nicknamed "the monster" for the vast number of elements it contained, was found by Robert L. Griess, Jr. of the University of Michigan. In what has been called the longest proof ever (although it has since been beaten by computer proofs), the bits and pieces of work with groups were put together to show that all the finite groups had been properly classified and there were no more around to add to the list. Headlines in science journals put it this way: "Mathematicians put themselves out of work." By 1980, this part of group theory was completely finished. You might compare this with the discovery of regular solids in Classical Greek times. Mathematicians had nothing more to do along this line after the icosahedron, a regular solid with 20 faces, was discovered; what is more, they could prove that the five known regular solids were all that there were ever going to be.

Scientists knew about group theory, but at first they did not know what, if anything, to do with it. Crystallographers discovered early on, however, that it could be used to classify crystals. Then chemists found uses for groups and, by the 1930s, physicists had found that group theory could help explain various kinds of problems. In this, they were influenced by Hermann Weyl, who brought Cartan's work into physics.

Group theory in particle physics was used as early as 1939, when there were not many particles known. Eugene Wigner had been working on some sensible way to classify the six known particles and their antiparticles. He was not getting anywhere, so he asked his old high-school buddy John von Neumann if von Neumann had any suggestions. Since Wigner's old high-school buddy was among the most brilliant mathematicians ever, it is not surprising that von Neumann pointed Wigner toward a theorem in group theory that solved Wigner's specific problem. As a result, Wigner discovered that elementary particles behave as if they are the elements of a mathematical group.

It is not surprising that elementary particles behave this way. Groups are very general, since the rules that describe them are so broad. Also, group theory can be thought of as "generalized symmetry," and physical laws for particles, such as conservation laws, derive from underlying symmetries. Finally, particle interactions can actually be pinned down as transformations. Some of these transformations are closely related to mathematical transformations of space, while others represent physical transformations; but, just as with purely mathematical transformations, transformations that leave some properties invariant can be arranged into a group. Thus, group theory turns out to be ideal for studying particle physics.

Wigner's scheme was an important first step, but further work with group theory did not immediately occur. When it did, it was largely inspired by two problems of the late 1950s—the need to understand the weak force better, especially since it did not conserve parity; and the need to classify all those new particles.

Group theory was certainly in the air around 1960. A large number of physicists, many working independently, began to use group theory in the early 1960s to obtain correct results in several different cages of the particle zoo. The first theories to be described are essentially the work of Murray Gell-Mann, Sheldon Glashow, Yuval Ne'eman, Abdus Salam, and Steven Weinberg.

The problem was still in understanding the weak force. Further work on parity showed that parity is not conserved because the weak force works only on left-handed particles and on right-handed antiparticles. For example, a left-handed electron can interact weakly with a left-handed neutrino or a right-handed positron (antielectron) can interact weakly with a right-handed antineutrino. Because electrons and positrons have a mass, albeit small, an electron or positron can be stopped and pointed in the other direction. When this happens, the handedness is changed. Parity conservation occurs with the right-handed electron and the left-handed positron. Neutrinos, however, do not change handedness. In the 1960s, it was assumed that the neutrino had no mass, which would explain why all neutrinos were left-handed and all antineutrinos were right-handed; since a massless particle travels at the speed of light, it cannot be stopped and reversed. Today, many physicists think that the neutrino has a small mass, but now they cannot explain why only left-handed neutrinos exist. Since the existence of only left-handed neutrinos fits experiment, we can ignore the question of the neutrino's possible mass in this discussion.

In analogy with the electromagnetic force, the weak force can be described in terms of a charge. The charge of the electron is -1, but the weak charge depends on whether the electron is left-handed or right-handed. A left-handed electron is assigned a weak charge of $-1/2$, while a right-handed electron is assigned a weak charge of 0. Since there is no right-handed neutrino, the right-handed electron cannot interact weakly with another particle, as the left-handed electron interacts with the left-handed neutrino. Similarly, the left-handed neutrino, which has an electromagnetic charge of 0, is assigned a weak charge of $+1/2$. In this way, the weak force between the left-handed electron and the neutrino is 1 (unit of weak charge, not to be confused with the much larger 1 unit of electromagnetic charge).

Physicists had developed a good theory of the electromagnetic force which seemed to do everything except explain its value—which remains unexplained today. In this theory, the force is carried by photons that are exchanged between particles. If two particles both have negative charges, this exchange of photons will tend to push them apart. Also, if the particles both have positive charges, the exchange will push them apart. When the particles have opposite charges, one positive and one negative, the exchange will tend to pull the particles together.

It was in analogy to this theory that Yukawa developed his theory of the strong force, which predicted that the particle we now call the pion would be exchanged between particles that feel the strong force. Unlike the exchange of photons, however, the pion does more than mediate a force. The arrival of the charged pion also changes the electromagnetic charge of the particle, switching a neutron into a proton, for example. But there are also neutral pions that carry the strong force and do not change the electromagnetic charge. Thus, two neutrons or two protons can interact strongly with each other.

One way to view the interchange of particles is as a transformation, similar to a mathematical transformation. The interchange of a photon between two electrons is an identity transformation; it does not change either electron. It is possible to have a group in which the only operation is the identity transformation. Such a group has the group property, is associative, and has inverses for all the elements. Specifically, there is a group in Cartan's classification scheme called U(1), for "unitary group 1," that has this property. When physicists applied group theory to quantum electrodynamics, the theory that explains the electromagnetic force, they found that the U(1) group can be associated with the electric field. By itself this was interesting, but not especially useful.

There is also another way of looking at a force, which is in terms of a *field*. In 1954, Frank Yang and Robert Mills took another idea of Hermann Weyl's and combined it with Noether's theorem to produce a special way of looking at fields. The basic idea is that there exists a compensating field that preserves symmetry; indeed, it arises from symmetry. In the case of the electromagnetic force, this compensating field, called a *local gauge field,* is what keeps the electric charge invariant. This part of the idea was worked out by Weyl and Fritz London. Yang and Mills showed that the same kind of mathematics could be used for other invariant properties. Specifically, they showed that a mathematical construct called *isotopic spin* was conserved by a local gauge field related to the strong force. Unlike the local gauge field for electromagnetism, their field was based on a noncommutative transformation. The Yang-Mills fields, as noncommutative local gauge fields came to be called, did not produce the same theory of the strong force as Yukawa's. Instead of pions, they predicted that a kind of boson, now called a *vector boson,* carried the strong force. They also predicted that these particles would be massless, like the photon. This last part was the most disturbing, because it should be easy to observe massless vector bosons, and none had ever been observed. The Yang-Mills theory, admired for the results that could be obtained by starting with pure symmetry, was left on the shelf in the 1950s.

Where there is invariance and symmetry, however, a group is generally lurking. A few physicists playing with Yang-Mills fields recognized that each such field would be associated with a definite group, and that for every group of particles that could be the transformations of other particles there would be a Yang-Mills field. In the case of the Yang-Mills field that produces electromagnetic charge, the group turned out to be U(1), as was also deduced from other evidence.

If a group could be associated with one force, however, it might be possible to associate a group with another force and get something more interesting. Physicists suspected that there might be a connection between the electromagnetic force and the weak force.

The Yang-Mills field for the weak force could be worked out on the basis of conservation of *something.* The trick is finding the quantity to be conserved, such as charge or isotopic spin. Several attempts were made to do this that were good enough to produce essentially correct results, even though their basis in fact was quite foggy. The weak force had to have the kind of a group associated with the transformations of a book by rotating it through right angles in 3-space. This is also the group of possible rotations of isotopic spin in isotopic-spin space, a mathematical space with an infinite number of dimensions. (Yang and Mills had the right idea in 1954, but they applied it to the wrong force.) This group was classified by Cartan as SU(2), for "special unitary group 2." In this case, the 2 refers to the two particles involved with the weak force, the electron and the neutrino. (The situation is more complicated than that, but these difficulties need not be discussed now.) SU(2) requires three vector bosons as the transformations. Unfortunately, as Yang and Mills had shown, the vector bosons should not have any mass, but that did not make any sense. No one had ever seen a charged particle that had no mass, and two of the three particles would have to have charge. Furthermore, one of the particles must be neutral with regard to charge. Interactions involving the neutral particle ought to be observable. Such interactions are called *weak neutral currents* by physicists. Since no one had observed any massless vector bosons nor any weak neutral currents, there must be something wrong with the theory.

Still, the theory was attractive for other reasons. Two groups can be combined by an operation called the *cross product,* which is usually indicated in group theory with a times sign. If you form the cross product of the group for the electromagnetic force with the group for the weak force, indicated by SU(2) × U(1), the result is a group that includes both forces. Thus, investigators searched for weak neutral currents in the hope that the theory would be correct after all. If so, the number of fundamental forces could be reduced by one, from four to three.

When the strong force was jammed into the same group, however, even more problems arose. Murray Gell-Mann was trying to explain the strong force using a Yang-Mills field, but was hampered by not knowing group theory. When Sheldon

Glashow showed Gell-Mann his model of SU(2) × U(1) in 1960, Gell-Mann thought that something similar would solve his problem with the strong force.

Gell-Mann almost became like the man in the old joke who was trying to invent a new soft drink. The inventor worked in his basement every night, inventing in succession 1-Up, 2-Up, 3-Up, 4-Up, 5-Up, and 6-Up. Then, discouraged, he gave up, not knowing . . .

Gell-Mann, who was in Paris at the time, would go to lunch with French friends, eat a lot, and drink a lot of French wine. Then he would return to his office to calculate. On the afternoon after he decided to try group theory, he worked his way through groups with three transformations, four, five, six, and seven. Nothing worked, and he was sleepy from all the food and wine. So, like the inventor in the basement, he gave up.

Months later he tried again, still getting nowhere. One day he described his struggle to Richard Block, an assistant professor of mathematics at Caltech who was an expert in group theory. Block pointed out that a group of eight transformations, known as SU(3), would do the job. It was also the simplest group that was not a cross product of SU(2) and U(1).

The SU(3) model fit some known facts about the strong force and the particles involved in strong interactions very well, and was a little vague about other facts and particles. In some places, new particles had to be predicted to make the fit. Because it used eight transformations, Gell-Mann nicknamed the model "the eightfold way," after one of the fundamental ideas of Buddhism, and published. The theory was much admired, but like the SU(2) × U(1) theory, it did not seem to fit the experimental facts. For one thing, it predicted too many unknown particles.

Other physicists also developed SU(3) models of the strong force around the same time, 1960. All had problems with the experimental data, but in 1962 the tide began to turn. New experiments began to confirm predictions of the eightfold way and some of the predicted particles were discovered to exist. Gell-Mann was emboldened to predict the specific mass and other properties of yet another particle. When it was found, after a couple of years of searching, the eightfold way was a certified hit. Although it had been preceded by the SU(2) × U(1) theory, the eightfold way was the first theory based on groups to be generally accepted by the physics community since Wigner's theory.

On the other hand, it was not clear at the physical level why the eightfold way should work. Another step was needed to make this purely mathematical construct meaningful to physicists. It was all well and good to say that symmetry would result in just this kind of a set of particles, but other symmetrical theories were around that did not work. For a time the eightfold way competed with a different SU(3) theory, which was discarded only because it did not fit with experiment. At bottom, physicists want some sort of physical reality.

One of the physicists seeking this reality was Robert Serber at Columbia University. Furthermore, he was aware of a defect in the eightfold way as it applied to group theory. A group should have a fundamental representation from which the other members of the group can be created. The SU(3) group, although it has 8 transformations, should have a fundamental representation with 3 members from which the transformations could produce the other members of the group. SU(3) as a group had to have those 3 simplest members, but the eightfold way as a theory was totally silent about them. Serber was having lunch with Gell-Mann, so he pointed out these deficiencies and asked Gell-Mann why he had not considered adding 3 *little* particles, which Serber called pieces, to the theory, to match the group requirements. That way, Serber said, you could take 3 of these little pieces and make a proton or a neutron; you might take a piece and an antipiece to make a meson.

Gell-Mann explained that he had not considered such a thing because the pieces would have to have a fractional charge, and all known particles had integer charges. But he thought about it some more, and the next day he began to work out what we now know as the *quark model.* He published a year later in February of 1964, just about the time that the eightfold way gained its respectability by a successful prediction. Even then, Gell-Mann was still skeptical about the fractional charges, especially since to get rid of them, he had to invent a fourth piece, which did not fit the group theory. Therefore he first explained that there might be four pieces, three of which he called *quarks* after a line in *Finnegans Wake* and one other that he called *b*, all with integral charges. Then, he showed that just the three quarks would do the job, provided they had charges of $+2/3$, $-1/3$, and $-1/3$.

Gell-Mann showed that with the fractional charges, Serber's original idea would work. He called the $+2/3$ quark an *up*, one $-1/3$ quark a *down*, and the other $-1/3$ quark *strange*. Thus, the proton could be made from two ups and a down, giving it a charge of 1, while the neutron could be one up and two downs, giving a charge of 0. Like all particles, quarks would come in pairs. For every quark, there would be an antiquark. Lighter particles, the mesons, could be a quark and an antiquark, say an up and an antidown for a charge of 1. Furthermore, the strange quark could be used to account for strangeness. The proton has no strange quark, so its strangeness is 0. The weak force might be able to change the "flavors" of quarks, so that weak forces could change a strange quark into an up or a down, which would account for the observed slow decay of strange particles.

Although no one had ever seen a quark, or anything like one, the theory explained so much that it was hard not to accept it. It was a triumph of group theory over physics. Some physicists were of the opinion that the quarks were just a mathematical trick that worked, but others started to work on ways to prove that the unseen quarks were real.

It should be noted that Serber and Gell-Mann were not alone in realizing that the SU(3) had to have a fundamental representation with 3 members. About a half-dozen other physicists came up with the idea at roughly the same time, but either could not make it work, could not explain why it worked, or could not get such a crazy idea published.

By 1968, the first evidence began to appear that quarks were real. In the meantime, it had become clear that there were problems with the quark model. It did not explain everything, and it did not really fit its own Yang-Mills field. Again, a number of physicists tried to fix the model. The result was *another* SU(3) theory, one that kept quarks but that had the original Gell-Mann SU(3) theory as a kind of consequence.

The new SU(3) theory was named (by Gell-Mann, as usual) *quantum chromodynamics.* It eventually came to have 6 quarks, each of which comes in 3 colors, confusingly designated in different ways by different physicists. Patriots favor red, white, and blue; while artists favor red, blue, and yellow; and iconoclasts prefer red, green, and blue, for example. We will use magenta, cyan, and yellow, the three colors of process printing. In quantum chromodynamics, quarks are held together by a color force, mediated by 8 particles called *gluons.* These are the vector bosons—spin 1—of SU(3), corresponding to the 3 vector bosons of SU(2) and the single vector boson of U(1). The name *gluon* is so intuitive for the particles that glue the universe together that some physicists have decided to call all the vector bosons *gluons,* but we will keep them separated for clarity's sake. Like the other vector bosons, emission and absorption of a gluon can cause a change in the particle or leave it alone. Six of the gluons change a quark's color, from magenta to cyan, magenta to yellow, cyan to magenta, and so forth. The other two are interchanged without producing a color change.

Experiments do not see the color for two reasons. The mathematics of the Yang-Mills field shows that color force works in a way opposite to more familiar forces; the farther apart two quarks are, the stronger the color force—so quarks always stay close together, although they can move around easily when they are very close. Also, colors in observable particles are always either combinations of all three, producing black, or of the three anticolors, producing black, or of one color and its anticolor, producing gray. In all observable particles, the color vanishes. Experiments can, however, see results that are best explained in terms of color. High-energy collisions of electrons with positrons produce enough energy to create a quark antiquark pair, each member of which moves in a different direction, decaying to form a characteristic pair of jets in a particle detector. The particles observed in the jets are those predicted by quantum chromodynamics.

In this SU(3) theory, the 3 entities needed for the fundamental representation are the 3 colors, not the three quarks—which is just as well, since there have turned out to be 6 quarks. Add charm, truth (or top), and beauty (or bottom) to up, down, and strange.

What became of the strong force and the pion interchange that was supposed to cause the strong force? Physicists now say that this is an indirect result of the *true* strong force, which is the same as the color force. The pion strong force can be likened to the van der Waals force that attracts molecules together as a result of fluctuations in the electromagnetic field. Similarly, the true strong force (color force) between quarks produces the pion strong force between particles composed of quarks.

Recall that protons and neutrons are composed of up and down quarks and antiquarks. In quantum particle physics, all particles are surrounded by clouds of virtual particles and antiparticles. In the case of the electron, the antiparticles, which are positrons, are attracted to the "bare" electron by the electromagnetic force. In the case of a proton or neutron, the part of the cloud nearest the particle will contain an appropriate mixture of up and down quarks and antiquarks, attracted by the color force from the "real" up and down quarks or antiquarks that make up the "bare" particle. If two nucleons (the generic term for proton or neutron) are close together, these virtual up and down quarks and antiquarks will combine in suitable combinations to form virtual particles. And what will these combinations be? An up quark combines with a down antiquark, which forms a positive pion. Similarly, an up antiquark combines readily with a down quark to form a negative pion. Finally, the up quark–antiquark pair or the down quark–antiquark pair produce the neutral pion. Thus, between two nucleons there is an apparent interchange of virtual pions, just as predicted in the Yukawa theory. The main difference in the quantum-chromodynamic view is that this is an effect, rather than a cause.

Quantum chromodynamics based on SU(3) was so successful that it seemed even more likely that SU(2) × U(1) ought to be also successful. A new concept needed to be added to the ideas of

Yang-Mills fields and group theory to make it work, however. This concept is explained in the next chapter.

Summary

In this chapter, we first looked at a group that emerged from the permutations—interchanges—of three objects to introduce the general concept of a group. Operations with a triangle revealed that groups are a formal and generalized form of symmetry; that is, groups often arise from symmetry operations. Then we saw that groups could be used to explain and predict the behavior of elementary subatomic particles. This occurs because the behavior of subatomic particles is largely dominated by conservation laws, and conservation laws directly arise from symmetries. Since groups also arise from symmetries, group theory has come to dominate particle physics, which many people believe is the most fundamental science of all.

9

Taking Symmetry to New Heights

We already see features in the theory that approximate the phenomena observed in the universe. There may be no fundamental obstacle to explaining nature with this theory; we just have to understand the mathematics better. We're not yet in a situation where experimentalists can shoot us down, because we don't yet know the predictions of the theory.

John Schwarz

For some minutes Alice stood without speaking, looking out in all directions over the country . . . "I declare it's marked out just like a large chessboard . . . all over the world—if this is the world at all."

Lewis Carroll

Although physicists have come more and more to rely on symmetry principles for explaining the universe, it is important to note that the universe is not perfectly symmetrical. If it were, everything would be just like everything else, and it is not, fortunately for us. The different kinds of quarks are not the same, for instance. If they were all exactly the same, nothing recognizable would exist. There is not even perfect mirror symmetry. In our part of the universe, there is much more matter than antimatter. If the amounts of matter and antimatter were equal, then everything would decay into pure energy.

Until the early 1960s, however, physicists took it for granted that some things did not look exactly like other things, and concentrated on the symmetric cases, those in which one thing looks just like another or like its mirror image. It was not that they did not know that the universe is filled with asymmetries along with symmetries. Instead, the thought had not occurred that the origin of most of these asymmetries might be interesting to study—an exception, of course, being the asymmetry between matter and antimatter.

Broken Symmetry

In some situations, a symmetry can change into an asymmetry, especially if the first symmetry is of the most generalized type. One example occurs with ferromagnetism. At sufficiently high temperatures, above 770°C for iron, there is no magnetism possible because the spins of the electrons in iron are aligned randomly. Since no direction is preferred, this situation is perfectly symmetrical in the same sense that a sphere is perfectly symmetrical. Even in the presence of a strong magnetic field, the thermal effects are strong enough to preserve this symmetry. As the temperature is lowered, however, the situation changes. Electrons in cooler iron that is in a magnetic field line up in pairs that gradually form tiny domains of magnetism. The initial symmetry is broken, for the magnetic field imposes a preferred direction for north and south poles. As the temperature is lowered, this preferred direction becomes even stronger, so that the greatest possible magnetism is at absolute zero. Even when the imposed magnetic field is turned off, the preferred direction remains, for the magnetic domains do not move completely back to their original positions. Observing the permanent magnet that results gives no hint of the original symmetry that existed at higher temperatures.

Another example of such a broken symmetry occurs when you apply pressure to the two ends of a cylindrical metal rod. Up to a certain pressure, nothing discernible happens to the rod. It maintains its symmetrical shape, a symmetry based on rotation around the axis of the cylinder. At sufficiently high pressure, however, this symmetry breaks and the rod bends. Now there is no longer any symmetry about the axis of the cylinder. Furthermore, as in the case of the magnet, this new state persists.

These effects were too well-known to be interesting, but a new phenomenon is always interesting. One such phenomenon, discovered in 1911, is superconductivity. At a sufficiently low temperature, mercury and lead conduct electricity with 0 resistance. This form of superconductivity was eventually explained, but the explanation troubled people because the explanation broke a symmetry found in electromagnetic theory (electrons that should repel all other electrons, when cooled sufficiently, began to form pairs). A few physicists began to get the idea that it was possible that symmetry breaking was a way of obtaining results that would not appear if everything were symmetrical.

Just as superconductivity appears when you break symmetry, broken symmetry might be able to explain other mysterious phenomena.

One of the physicists inspired by this possibility was Jeffrey Goldstone of Cambridge University. In 1960, Goldstone applied this idea to particle physics. He assumed that that all space contained a special symmetry-breaking field that could account for the difference in mass between an electron and a muon. When he worked out the mathematics for this, he was surprised to find that the new field also would require a new particle with zero mass. As in many different theories of the 1960s, such a particle was an unwelcome visitor; if zero-mass particles existed, people should be finding them everywhere, and they were not. Despite this problem and despite the theory being mostly speculation, Goldstone published. Within short order, Salam and Weinberg, who had discussed the idea extensively with Goldstone, rigorously proved in three different ways that Goldstone's idea had to be correct; that is, if there were a symmetry-breaking field, there also had to be new massless particles to go with it.

Goldstone's starting place had been the broken symmetry of the electromagnetic field that occurs with superconductivity. A condensed-matter (formerly known as solid-state) physicist from Bell Laboratories, Philip Anderson, pointed out that Goldstone, Weinberg, and Salam had to have made a mistake somewhere, since superconductivity is a broken symmetry *without* massless particles. His notion was that the embarrassing Goldstone particles and the equally embarrassing zero-mass particles of the SU(2) × U(1) theory probably cancelled each other out, resulting in a particle with mass. In short order, the Scot Peter Higgs and several other physicists showed that for Yang-Mills fields—and for Yang-Mills fields only—Anderson's idea was correct. The two kinds of massless particles got tangled up with each other and produced particles with mass. Along the way, another mysterious particle, now called a Higgs particle, showed up, however. Actually, four Higgs particles appeared, but three of them then disappeared in the process of giving mass to other particles. The fourth Higgs particle stayed around. It had the peculiar attribute that it took energy to make it go away, so that there are always Higgs particles in any vacuum.

At this point, all the machinery was in place for a major breakthrough, but for a while nothing happened. One of the reasons nothing happened was an old error repeated—physicists tried to apply the theory to the wrong problem, in this case, to the strong force instead of the weak and electromagnetic forces. In 1967, however, Weinberg suddenly realized that he had the solution to SU(2) × U(1) right before him. Despite having the right answer, Weinberg's paper explaining it fell on deaf ears. There were so many crazy theories floating around at the time; this was just one more that no one could prove right or wrong.

There was a major problem with the theory. Although it no longer predicted particles that couldn't exist, it still predicted interactions that had not been observed. These *weak neutral currents* were in principle observable. Experimenters had looked for them, however, and failed to find them. A new series of particle detectors came into use around 1972. Several experiments were conducted to look again for the weak neutral currents using the new equipment. These new experiments found the predicted interactions at about the predicted frequency. By 1974, it appeared that SU(2) × U(1) was the winner out of all the crazy theories of the 1960s.

At this point, the SU(3) theory for strong interactions was already well-accepted, so it was logical to combine the newly vindicated SU(2) × U(1) theory with that one, producing SU(3) × SU(2) × U(1). Other experiments continued to show predicted effects. Finally, in 1983, a team led by Carlo Rubbia cleverly modified a particle accelerator to be powerful enough to create the massive particles predicted by SU(2) × U(1). The theory was convincingly confirmed, and the combined SU(3) × SU(2) × U(1) theory came to be the *standard model* of particle physics. Particles do come in groups. Symmetry breaking is important. Almost everyone connected with the development of the theory and the experimental proof won a Nobel Prize (over a period from

1969 through 1984 including: Murray Gell-Mann, John Bardeen, Leon N. Cooper, John R. Schrieffer, Philip W. Anderson, Steven Weinberg, Sheldon L. Glashow, Abdus Salam, James W. Cronin, Val L. Fitch, Carlo Rubbia, and Simon van der Meere).

It Takes GUTs to Predict the Disintegration of Matter

Since group theory combined with broken symmetry works so well in the standard model, it seemed very likely to various theoreticians that with a little tinkering it might be possible to go beyond the standard model. Just as it seemed awkward to have four forces and hundreds of particles, it seemed awkward to have three different groups. Furthermore, $SU(2) \times U(1)$ actually showed that two of the forces were different aspects of the same force. Above a certain energy, before the symmetry is broken, electromagnetism and the weak force are the same—a single force called the *electroweak* replaces the two separate forces. Perhaps there could be even more symmetry at a still higher force, perhaps even enough symmetry to unite the strong force with the electroweak force. Physicists began to work on ways to unify the forces.

Efforts to unify forces, other than the electromagnetic and weak, have a history marked by spectacular failure. Recall that Einstein worked for about 30 years on a program to unify the electromagnetic force with gravity, and he failed. Physicists suspected that perhaps the main difficulty Einstein faced came from gravity. If the strong and electroweak forces could be unified, ignoring gravity, then perhaps it might be possible to work in gravity at some later time.

The problems in unifying forces were many, even ignoring gravity. For one thing, the strong force did not have any effect on the light particles of the electron family—known as *leptons*. The leptons consist of the electron and the two "heavy" electrons, the muon and tauon, as well as a neutrino for each of the kinds of electrons. The particles that are affected by the strong force are called *hadrons,* which come in two groups: the *baryons* that consist of three quarks and the *mesons* that are made from two quarks. Leptons seemed to have nothing to do with quarks. *Bosons,* the other major group of particles, are the particles that mediate the forces. The set of bosons for the electroweak force is totally different from the set that mediates the strong force.

The first effort to unify most of the particles and three of the forces came in 1973 when Abdus Salam and Jogesh Pati tried to put the quarks and leptons on an equal footing. They began by saying the the electron was actually a fourth color quark, which they dubbed lilac. This theory did not seem to many people to make much sense, although it produced one remarkable prediction. It is the essence of group theory that elements of the group can be transformed into each other. Thus, if you put quarks and leptons into the same group, it must be possible to change a quark into a lepton. This transformation could also effectively change a baryon into a meson, since you would now have a lepton, unaffected by the strong force, and two quarks, although the quarks would only form a meson by a proper choice of transformation. In any case, the baryon would no longer exist, and the meson would decay into pure energy.

Most baryons, such as the neutron and lambda particle, decay anyway, so this is not an obvious problem for these particles. When they decay, however, the end product of the process always contains another baryon, so the total number of baryons appears to be conserved. Furthermore, the baryon that is most important to the existence of the universe, the proton, has never been unequivocally observed to decay. The conventional explanation of why the proton is not observed to decay is a conservation law called conservation of baryon number. The conservation of baryon number was suspect, however, because no one knew what symmetry could give rise to it. According to Noether's theorem, a conservation law only occurs as a result of symmetry. Thus, although physicists used the conservation of baryon number, a few, such as Andrei Sarkharov, had suggested that perhaps it did not always work. This uncertainty about conservation of baryon number encouraged Salam and Pati.

The Salam-Pati theory said that baryon number should not be conserved. For all practical purposes this is the same as saying that protons must decay some of the time. It was easy to see that this could not happen very often. For example, human beings contain vast numbers of protons. Suppose only one proton decayed every 10^{16} years (in scientific notation, 1 followed by 16 zeros is 10^{16}.) Our bodies would be so radioactive that we would die. The fact that we are alive sets a lower limit on how often a proton might decay. The Salam-Pati theory predicted that one proton would not decay for 10^{28} years at least. This was good news because atoms are made from protons, and if all the protons decayed, there would be no elements left in the universe.

Frederick Reines had already experimentally measured the possibility of proton decay. His experiments showed that protons did not decay in less than 10^{27} years. Thus, the Salam-Pati number of 10^{28}, which is ten times as long as Reines' lower limit, looked safe.

You can't measure proton decay by looking at a single proton and waiting 10^{27} years for something to happen. Most astronomers think that the lifetime of the universe is at most 20 billion years old—less than 10^{11} years in other words. To measure proton decay, you get 10^{27} or more protons together in one place and expect one of them to decay in the course of a year. Particle decay, like the other effects of quantum physics, is ruled by probability, and the probability of a single proton out of 10^{27} decaying in a year is the same as the probability that a single proton chosen at random will decay in 10^{27} years. Better yet, get 10^{30} protons together in one place, and if they decay at the predicted rate of 10^{28} years, then you should see a hundred decays among your 10^{30} protons. Each proton has a very small mass, roughly the same as a hydrogen atom. According to chemical theory, at standard temperature and pressure, 22.4 liters of hydrogen gas contains 1.204×10^{24} protons. Therefore, 22,400,000 liters of hydrogen gas would contain somewhat more than 10^{30} protons. In practice, however, it is more practical to use the protons in water, since each molecule of water contains 10 protons, or in iron, which has 26 protons per atom. If you get three hundred tons of iron together in one place (which is being used as a detector in one proton-decay experiment), you have a lot of protons to watch.

About the same time as the Salam-Pati theory was announced, Sheldon Glashow and Howard Georgi tried another approach to putting the leptons and quarks into the same group. They looked for a group that was big enough to contain SU(3), SU(2), and U(1) as subgroups. Working with Cartan's classification scheme, Glashow and Georgi showed that only nine groups were suitable to contain SU(3), SU(2), and U(1) as subgroups and that of these nine, only SU(5) seemed to match reality. They published their results early in 1974. No one was much interested in another group at the time. The SU(3) quark theory was just becoming commonplace, and the evidence for SU(2) × U(1) had just been found but not announced as yet.

Later in the year, however, Georgi collaborated with Helen Quinn and Weinberg on calculating further implications of SU(5). They were able to show that the strong force and the electroweak force are identical in strength only for very small distances—much less than the diameter of a proton—or, equivalently, for very high energies. Specifically, the energy level at which the electroweak and strong forces merged was predicted to be 10^{14} GeV. Another way to look at this is that this energy can be measured at 10^{16} degrees Celsius, a temperature not known in the universe since shortly after the Big Bang that started it. (Sadly, temperature is a poorly understood concept, even by physicists. Heat is a form of energy, so GeV can be directly translated to joules, the standard macromeasure of energy, or to calories, the standard macromeasure of heat. Temperature cannot be directly translated. For example, the standard definition of a calorie is the amount of heat needed to raise a gram of water one degree Celsius. Therefore, while temperature is related to energy—as in the calorie definition—it is not the same. Nevertheless, physicists continue to express energies in terms of temperature.)

Thus, we have another example of broken symmetry. The symmetry between the two forces was perfect for a tiny fraction of a second after the Big Bang, and then was broken forever more.

Furthermore, Georgi, Quinn, and Weinberg were also able to calculate the rate of proton decay predicted by SU(5). It was 10^{32} years, safely long enough so that proton decay was not a problem for the theory.

What was a problem for the theory was that it predicted effects that were impossible or difficult to find and it also predicted some wrong numbers for quantities that had already been measured. This did not deter people from continuing to work on the idea, both theoretically and experimentally. Some theoreticians looked at other groups that might also unify the forces. Eventually, various theories that unified the strong and the electroweak forces were produced. All such theories are classified under the common name of Grand Unified Theories, or GUTs. These GUTs have a number of features in common, but use different groups as their basis and make slightly different predictions. Other theorists tried to improve the calculations based on the SU(5) GUT. Some experimentalists carefully remeasured quantities whose values were different from those predicted by the GUTs. The new calculations and the new measurements seemed to converge toward the same numbers, giving a new credibility to GUTs.

Most physicists do not think that SU(5) as it is currently understood is the right GUT, but it remains the simplest, and other more complicated GUTs predict in a general way the same things. Even with the doubts about its ultimate validity, it is easiest and most productive to take SU(5) as the example for all GUTs.

SU(5) is a group that consists of transformations of five distinct objects. In Glashow's and Georgi's version, the five objects are taken to be all right-handed. Specifically, they used the right-handed versions of the magenta, cyan, and yellow down quarks, a right-handed positron, and the antineutrino (which is always right-handed). To get all the possible transformations of these particles, they need 24 bosons (Figure 9–1). The first 12 of these are the various gluons familiar from SU(3) × SU(2) × U(1): the photon, or the gluon of the electromagnetic force; the three intermediate vector bosons that are the gluons of the weak force; and the eight gluons that create the strong force (also known as

	magenta (right)	cyan (right)	yellow (right)	positron (right)	antineutrino
magenta (right)	I (2 colorless gluons, photon, Z°)	Gluon (magenta to cyan)	Gluon (magenta to yellow)	X	X
cyan (right)	Gluon (cyan to magenta)	I (2 colorless gluons, photon, Z°)	Gluon (cyan to yellow)	X	X
yellow (right)	Gluon (yellow to magenta)	Gluon (yellow to cyan)	I (2 colorless gluons, photon, Z°)	X	X
positron (right)	X	X	X	photon + Z°	W^+
antineutrino	X	X	X	W^-	Z°

FIGURE 9–1
The twenty-four different particles of a popular GUT. Note that the dozen X particles, which have never been observed, are supposed to be different from each other.

the color force in this context). The remaining 12 transforming particles have never been observed; they are all designated as X. The dozen X particles are the ones that transform leptons into quarks and vice versa. The reason that X particles have never been observed is that they are very massive, so they could only be created by the high energy available during the Big Bang. After that, they all decayed.

You may wonder why, if there are no X particles around, theorists still think that it is possible for protons now existing to decay. Such a decay would require an X particle to exist, so that it could transform a quark into a lepton. As before, it is a matter of probability. The vacuum is a seething mass of virtual particles of different energy levels. You may have heard of the idea that if you wait long enough, all the molecules in a brick will be moving in the same direction at the same time, so the brick will suddenly jump into the air. The probability of the energy of a virtual particle being great enough to be an X is on this order. To test the brick theory, you get an enormous number of bricks and watch for the one that jumps up. To test SU(5) you get an enormous number of quarks (three to a proton) and watch for one of them to become a lepton. In other words, although there is not enough energy to create new X particles, sometimes it happens anyway. The law of conservation of mass-energy combined with the uncertainty principle guarantees that the X particle will not be around very long, but sometimes it will be around long enough to do its job.

By combining particles, you can produce electrons and all the other quarks and neutrinos. When you count up all the variations of left- and right-handedness that are allowed for electrons, neutrinos, and quarks and then throw in the antiparticles for all of these, you get 30 particles in all. The possible combinations of the 5 fundamental particles are just those 30 particles.

The 30 particles obtained in this way are those that make up ordinary matter (and ordinary antimatter!). At a higher energy than ordinary matter, you run into muons, the muon neutrino that is somehow subtly different from the electron neutrino, and the strange and charmed quarks. In SU(5) you can get all these particles by starting all over with another 5 fundamental particles, producing a second family of matter in this way. There is still the problem of Rabi's famous "Who ordered *that?*" Or why a second family should exist at all, but since one does, it is nicely described by SU(5). Furthermore, at a still higher energy level there are 30 more particles based on the tauon, its neutrino, and the truth and beauty quarks. So, by applying SU(5) three times, you can account for the 90 known members of the lepton and quark families, which in turn account for all the leptons, baryons, and mesons. Along the way, all the known bosons fit in neatly. Of course, you have to put up with the unobserved X particles.

The X particles are not so hard to take, however. SU(5) has, as one of its possible consequences, that interactions involving X particles violate time symmetry in the same sense as the neutral kaon decay violates time symmetry. Thus, it is possible that X particles could create more quarks than antiquarks. Most GUTs, including SU(5), assume that this is indeed the case. If so, at the Big Bang, when X particles had real existences instead of virtual ones, more matter than antimatter would have been created. If the same amount of matter and antimatter had been created, the universe would have fallen into a hole, resulting in all radiation and no matter. Thus, one of the consequences of the X particles is the existence of the universe.

Another consequence of the X particles is that a virtual X particle, arising spontaneously in the nucleus of an atom or nearby, will change a quark in the nucleus to a lepton, resulting in the decay of a proton. Eventually, this will happen to all the protons in the universe (if something else doesn't get them first; see below), and matter will cease to exist. So if the X particles produced the matter in the universe, they also will eliminate it. I guess that's a form of symmetry, too.

GUTs also invariably predict the existence of *magnetic monopoles,* particles that are like magnets with only a north pole or a south pole. A magnetic monopole is like a string with only one end, so it is a bit hard to imagine. As one approaches a magnetic monopole, its powerful, but odd, magnetic field affects the vacuum, causing virtual particles to rise out of the vacuum in great numbers. Nearer the center of the magnetic monopole, the field begins to restore broken symmetries. As you reach the edge, the electromagnetic and weak forces are reunited. Closer still, and the electroweak force becomes the same thing as the strong force. At the center of the monopole, there is perfect symmetry.

One result of the symmetry inside a monopole is that at the center leptons and quarks are the same kind of particles. Thus, a magnetic monopole should catalyze proton decay, just like a virtual X particle would, because one or more of the quarks that make up the proton before the monopole arrives could just as easily be leptons by the time the monopole passes. With perfect symmetry, there would be a fifty-fifty chance for each of the three quarks in a proton to turn into a lepton, which means that about 9 out of 10 times that a monopole encounters a proton, the proton will decay. A monopole passing through a detector for proton decay would leave a string of proton decays that would be a spectacular event. Since this has never been detected, monopoles must be rare if they exist at all.

Although there have been a couple of reports of monopole detection (by other means), these have been too infrequent to judge whether or not the monopoles were really present. When an experiment runs for several years and apparently detects a single monopole, many scientists doubt even the one example. Theorists have tried to patch up the prediction of magnetic monopoles in GUTs by calling on neutron stars, black holes, and unknown phenomena to eat the monopoles. If most of the monopoles are eaten, then they no longer pose a problem for GUTs.

GUTs make a number of other predictions that are intriguing and also difficult to prove or disprove. In addition to baryon number not being conserved, lepton number is also nonconserved. One consequence of this is that neutrinos may also behave like neutral kaons, which would give them a slight mass. Cosmologists believe that the universe must have more mass in it than has been observed. If neutrinos have a mass, even if very slight, they could account for this mass since there are lots and lots of neutrinos around. Furthermore, one form of neutrino could, under certain conditions, change from an electron-type to a muon-type or vice versa. Such a change has been suggested to account for the number of neutrinos produced by the sun being many fewer than what ordinary (non-GUT) theory predicts.

Supersymmetry and Other Super Stuff

Although GUTs are generally accepted, if not proved and not stabilized into a single theory, the success of symmetry ideas in physics has led to a string of new theories that go beyond the GUTs and offer different explanations for the fundamental nature of the universe. Heinz Pagels in his book *Perfect Symmetry* labeled these "wild ideas." Niels Bohr once remarked of an earlier wild idea about quantum theory that it could not be true because it wasn't crazy enough. Possibly one of these wild ideas is crazy enough to be true, but no one is sure yet and each idea has its champions and detractors. A glance at the index for a recent year of *Physical Review Letters,* where most of the wild ideas—if they are not *too* wild—get published, shows dozens of competing notions. Among these are the various "super" theories, such as *supersymmetry* and *supergravity.* Another group of theories involves either 10-dimensional or 11-dimensional space. The wild ideas don't exist in isolation, either. Supersymmetry can be combined with an 11-dimensional space, for instance. The reason that these theories are "wild" is that, so far at least, no one has been able to prove whether they are true or not. Physicists who object to the new theories feel that the proponents of the theories are like the medieval philosophers who raised complex questions about angels whose existence could not be demonstrated.

The grandfather of all is supersymmetry. The basic notion of supersymmetry is like the unwelcome conclusion of the GUTs that a quark can become a lepton. In supersymmetry, a fermion can become a boson. Recall that particles can be even or odd, in the sense of Chapter 4. An odd particle has a spin of $1/2$, which causes it to obey the Pauli Exclusion Principle. Such particles are called fermions, and include the leptons and the baryons. Other particles are even, since they have a spin that is an integer, which causes them to want to huddle together. These are the bosons, and they include the pion, the photon, other gluons, and the (unobserved) graviton. Victor Weisskopf has pointed out that fermions are really matter, while bosons are something else. I find it much easier, most of the time, to think of fermions as particles and bosons as waves (for example, bosons, such as photons, can get together to form giant radiowaves, but fermions, such as electrons, retain their identity even as they are producing images on a cathode ray tube). Yet another way of looking at this difference is to associate bosons with *force,* since they mediate the known forces, and to associate fermions with *mass,* since they form ordinary matter. In any case, it is clear that fermions and bosons are even more different in some senses than leptons and quarks. Thus, supersymmetry has more strange predictions than even the GUTs.

Why the name *supersymmetry?* A GUT such as SU(5) has more symmetry than SU(3) × SU(2) × U(1) because it allows an additional transformation, the transformation of a quark into a lepton or vice versa. Supersymmetry has more symmetry than SU(5) because it allows an additional transformation, of a fermion into a boson, and vice versa. Supersymmetry need not replace SU(5), however. A commonly used theory these days is called supersymmetric SU(5). In general, supersymmetry is usually added to one of the GUTs, producing a general class that have been nicknamed the SUSY GUTs (SUperSYmmetric Grand Unified Theories).

The supersymmetric transformation of a boson into a fermion should not be confused with the energy of a boson creating a pair of fermions, as happens when a photon generates an electron-positron pair. In that case, the spin 1 photon creates two spin 1/2 particles, so spin is conserved. The transformation permitted by supersymmetry would violate conservation of spin.

Supersymmetry taken pure makes unacceptable predictions, notably that the fermions and bosons that can turn into each other must have exactly the same mass. This is really awkward, since most fermions and bosons generally have quite different masses. It is true that the mass of the muon (a fermion) and the pion (a boson) are so close that they were confused with each other for 10 years, but the masses are not exactly the same. Thus, supersymmetry must be broken, just as electroweak theory requires a broken symmetry to give different masses for the W and Z particles from the photon. No one is sure exactly what breaks the symmetry in electroweak theory, and scientists are even less sure about what could break supersymmetry—but since symmetry breaking is not very well understood, anything is possible.

The result of broken symmetry is a whole new set of particles that have different masses from any observed particles. Of necessity, the masses have to be large or someone would have seen one. In theory, any particle with a small enough mass can be produced in a particle accelerator, getting mass from energy by $E = mc^2$. We build larger and more powerful particle accelerators in hopes of seeing more massive particles—and it works. Carlo Rubbia and his team were able to create the predicted massive W and Z particles by getting more energy out of a particle accelerator, for example. The predicted new supersymmetric particles must have masses so great that current particle accelerators do not have the energy to create them.

Thus, supersymmetry posits a particle partner for each of the known particles. The supersymmetry transformation subtracts $1/2$ unit of spin, so that an electron with spin $1/2$ changes to a boson of spin 0, while the photon with spin 1 has a superpartner that is a fermion with spin $1/2$. In the case of a particle with 0 spin, $1/2$ is added to instead of subtracted from the spin to get the superpartner. The new particles are referred to by adding the prefix "s-" or the suffix "-ino" to a form of the name of the familiar particle. The s- prefix is used for the superpartners of the fermions, producing *squarks, sleptons,* and *sneutrinos.* The -ino suffix is used with the superpartners of the bosons, producing *photinos, gravitinos,* and *gluinos,* for example. This results in the partner of the W particle being a *wino,* but who cares? Note that there are no "sprotons," "sneutrons," or "spions," since these particles are viewed as composites of quarks, but the slepton partner of the electron, the *selectron* is an important part of supersymmetric theory.

In each case, the particle starting with s- or ending in -ino is somewhat like the familiar particle, but if the original was a fermion, the partner is a boson, and vice versa; also, the "superparticle" has a lot more mass, at least enough not to be observed. Before the symmetry was broken, the particles and the superparticles were all one.

As noted, the superpartners have not been found, in most cases because they have so much mass that they cannot be created in particle accelerators. Another possibility also exists, especially for the photino. It might have a mass small enough to exist all about us, but we just don't know how to see it.

Supersymmetric theory includes the idea that there must be one superparticle that is stable, simply because it has the least mass of any of the superparticles. This stability arises from the same mechanism that keeps the electron around, since charge is conserved and it is the smallest charged particle. Most supersymmetric theorists think that the permanent superparticle is the photino, but recent evidence suggests that it might be a particle called the higgsino, which is the superpartner of the unobserved Higgs particle, the hero of symmetry breaking in the electroweak theory. In any case, the particle must interact weakly with ordinary matter. Great efforts had to be made to detect the neutrino, since neutrinos interact so weakly with other particles, but we are sure that streams of neutrinos are passing through the Earth (and our bodies) at all times. If the photino—or any of the superparticles—interacts as weakly as the neutrino, we would not see it no matter what its mass is.

This opens another possibility for the missing mass in the universe. If there are a lot of superparticles in the universe, they would supply the mass even though we could not detect them.

It would be possible to detect a superparticle in the same way that other elusive particles have been detected, which is by finding the right amount of energy missing in some reaction. Supersymmetry enables physicists to calculate various reactions that ought to produce superparticles. One such reaction could occur in collisions between electrons and positrons. While normally, an electron meeting a positron drops into the hole and produces a photon (gamma ray), at sufficiently high energies, supersymmetry predicts that a pair of oppositely charged selectrons would be produced. These would almost instantly decay into the original electron-positron pair plus a pair of photinos. Although we could not observe the photinos, the electron-positron pair would have lost roughly half their energy; furthermore, the directions of motion would be changed from what would be expected if the electron-positron pair had just been created from a random photon (which happens a lot at sufficiently high energies). Therefore, indirect evidence of the photino would be the sudden appearance in an electron-positron collider of pairs of low energy electrons and positrons going off every which way. Experiments of this type have been conducted, however, and have revealed no photinos. It is possible that the photino has too much mass to have been found in the recent experiments. New electron-positron colliders are being built and may prove that supersymmetry is actual or real.

Supersymmetry was soon generalized to include gravitation, producing a set of theories known as supergravity. The basic notion is to apply the supersymmetry transformation to Einstein's general theory of relativity. The general theory, while it has had many experimental verifications, is mathematically intractable. Miraculously, the supersymmetry transformation solves many of the mathematical difficulties, although the resulting mathematics is still so formidable that it keeps scores of theorists and computers busy trying to get some testable results.

One result of the general theory is that the force of gravity, like the other three forces in nature, must be mediated by a particle (wave). This particle, the graviton, has never been detected, but it is possible to calculate its properties. Since gravity extends to infinity, the graviton must have a mass of 0. Furthermore, it must have a spin of 2, which is also theoretically the largest possible particle spin—at least in most theories. The spin must be 2 as a result of the apparent absence of negative mass; if negative mass exists, the way negative and positive charges both exist, the graviton has a spin of 1, like the photon that mediates the electromagnetic force.

Applying the supersymmetry transformation to the graviton gives the simplest form of supergravity. It predicts the existence of the gravitino, a particle with spin $3/2$. Just as the graviton has never been observed, no particle with spin $3/2$ has ever been found and, until supergravity, even predicted. The resulting supergravity theory also produces local gauge Yang-Mills fields, which makes it fit better with SU(3) × SU(2) × U(1) than supersymmetry without gravity, which is based on a different kind of gauge field. Supergravity arose, in fact, from the conversion of supersymmetry to a Yang-Mills field, which produced results that clearly applied to gravity as well as to other particles.

This simple form of supergravity could not be reality, however well it works mathematically. It fails to link up the fields for gravity with the fields associated with other particles. It can be shown that there are only a finite number of theories that use local gauge fields to produce supergravity. The finite number in this case is 8. The simplest form is called $N = 1$, because there is only one supersymmetric transformation, and the other supergravity theories are known among physicists as $N = 2$, $N = 3$, . . . , $N = 8$. Most hope is held for the $N = 8$ theory because it produces a lot of fields, 163 of them in all. Each field is associated with, or perhaps the basic reality of, a particle. Applying the 8 transformations to the basic group produces 70 particles with spin 0, 56 with spin $1/2$, 28 with spin 1, 8 with spin $3/2$, and the graviton with spin 2. Unfortunately, the theory does not reveal which particle is which. If one of those spin 0 particles is the pion, one of the spin $1/2$ particles is the electron, and one of the spin 1 particles is the photon, for example, $N = 8$ supergravity is the theory of everything, the ultimate unified theory of the universe. Such a theory is called a Total Unified Theory (TUT). Some physicists think $N = 8$ supergravity is a TUT, but others think it is just elegant mathematics.

One of the reasons for skepticism is that supergravity does not predict anything special that experimentalists can check. Electroweak theory suffered from this dilemma until predicted neutral weak currents were found, and really became dogma when predicted W and Z particles were produced. GUTs still suffer from the apparent refusal of the proton to decay in an unequivocal way; that is, a few proton decays have been reported, but the physics community at large is yet to be convinced.

What Does It All Mean

The History of the Universe (Part 1)

In the beginning there was perfect symmetry.

Perfect symmetry means that there is no up or down, front or back, right or left. There is only one kind of particle and one kind of force. Perhaps there is only one particle and no force, or one force and no particle. Alternatively, there is chaos, an idea somewhat familiar from the Bible. If there is chaos, it is still impossible to recognize any direction. Perfect symmetry can also be found in perfect disorder.

There was no time. After all, this is the beginning, so it must be the beginning of time.

There was also perfect supersymmetry, if supersymmetry exists at all.

Then symmetry started breaking right and left. All sorts of things happened.

* * *

When particle physicists began to develop GUTs and supersymmetric theories, they also began to realize that testing these theories with any conceivable particle accelerators might be impossible. In particular, the GUTs implied that there was a factor of 10^{13} between the physics we could observe on Earth and the physics the theory predicted. This 10,000,000,000,000 fudge factor in the universe was very disturbing to many physicists. Supersymmetric theories, on the other hand, are popular in part because they predict particles that might be observed in accelerators being built or planned today. It was clear that the high energies needed to verify GUTs could only have existed around the time of the Big Bang. So some particle physicists became part-time cosmologists. This trend has continued, so that today most theoretical particle physicists routinely project their ideas back to the Big Bang as a way of checking them.

While details vary somewhat in different scenarios, there is remarkable consistency in the overall picture. Most physicists think we know a lot about the Big Bang and its immediate consequences. As they learn more, they hope to be able to predict the observable universe from first principles. There are still major problems to be resolved. Nevertheless, there is a set of remarkably clear ideas that have a lot going for them. They incorporate a combination of GUTs, supersymmetry, supergravity, symmetry breaking, SU(3) × SU(2) × U(1), and extra unseen dimensions into one story. Furthermore, in a path that goes beyond where we will follow, particle physicists proceeded to explain the creation of the elements in the observed frequencies, the life and death of stars and galaxies, the distribution of matter, and the fate of the universe. Along the way, they developed the theory of the inflationary universe, which is definitely worth a side trip. More and more theoretical physicists believe that this can all be put together into a single theory of everything. Only a few hold out from this wholesale leap into speculation, generally on the grounds that if you can't test it, it is not important.

Allen H. Guth invented the inflationary universe in 1979. Although Guth was a particle physicist, he began to worry over what the universe might have been like at the very beginning. He was motivated by the realization that the energy needed to deal with the most dramatic predictions of GUTs was only available before 10^{-35} seconds after the Big Bang. He considered what would happen if the very early universe reached a temperature below the one predicted for symmetry breaking in the GUTs. Note first that in cosmology, it is most intuitive to think of temperature as a measure of energy. A temperature of 10^{-14} GeV (giga electron volts) is the region in which strong force is still united with the electroweak force. If the temperature had ever gotten below the point where symmetry should break, and it had not broken, the matter in the universe would have been supercooled.

Supercooling is a well-known phenomenon. Water, for example, can be cooled far below 0°C, the normal temperature for freezing, and remain a liquid. Liquid water and ice are just two different *phases* of the same essential substance. (At some point in the fairly recent past, scientists stopped talking about "states" of matter and started talking about "phases" of matter.) As a liquid, water has more symmetry than ice, for ice crystals orient themselves in a form that, for ordinary ice, has period 6 symmetry, while water has period infinity symmetry. The period of liquid water is infinite because any rotation of liquid water does not change it. Ice, such as a typical snowflake, comes in hegagonal crystals that can be rotated into themselves 6 times. If water is supercooled, it retains its period infinity symmetry by staying a liquid. Toss a snowflake into sufficiently supercooled water, however, and the symmetry breaks. Suddenly all the supercooled molecules want to line up with the snowflake. The water quickly freezes into a structure with period 6 symmetry, a lot less than period infinity symmetry.

Guth worked out the mathematics of a supercooled vacuum that might exist at the beginning of the universe. Like supercooled water, the supercooled vacuum could undergo a sudden phase change. The peculiar equation of state for this phase change caused the universe to expand exponentially; in other words, at a rate that increased with increasing size. Such an expansion produces a comparatively big universe out of almost nothing in a very short time, which is called the *inflationary universe.* This original inflation should not be confused with the present-day expansion of the universe, which is approximately linear, not exponential; that is, the rate of expansion today does not seem to be increasing with the size of the universe; or, if the rate of expansion is changing, it is not changing by very much. It is even likely that the rate of expansion is slowly decreasing today.

Guth's original version of the inflationary universe excited cosmologists because it solved several major problems that were facing cosmology all at one stroke. When Guth, a particle physicist, first had the idea of the inflationary universe, he was unaware of all the questions his idea would resolve. As he got deeper into cosmology and learned of the problems, he began to realize what a good idea he had. Like Dirac, he had discovered that his equation was more intelligent than he was.

The first problem cosmologists faced at the time the inflationary universe was propounded was the lack of observed magnetic monopoles. According to the GUTs, there should be monopoles everywhere, but careful experiments revealed only a couple of candidate events, implying that monopoles were either very rare or perhaps even nonexistent. The GUTs claimed that when SU(5) symmetry broke to become SU(3) × SU(2) × U(1), magnetic monopoles would have been formed in abundance. Although no more would be formed after symmetry breaking, the magnetic monopoles (unlike the X particles) would not decay. After all, what could a magnetic monopole decay into? As noted earlier, it is believed that magnetic monopoles would catalyze proton decay, so a universe filled with magnetic monopoles would not stay a universe for very long. Thus, if you wanted to keep the GUTs, you had to get rid of the monopoles. It should also be noted that supersymmetric GUTs have just as many monopoles as ordinary GUTs.

Inflation to the rescue. Magnetic monopoles appear in GUTs when the symmetry of the electroweak force breaks. As the electromagnetic force appears for the first time, it appears with a different orientation in different parts of the universe. Where these fields of different orientation meet, twists in the field develop, some right-handed and some left-handed. These twists are the monopoles. If a right-handed monopole meets a left-handed one, one falls into the hole of the other, for one is the antiparticle of the other. If the electromagnetic force appeared after the inflation—in the Big Bang era that we still live in—the universe would be too big for very many of the monopoles to meet and annihilate each other. On the other hand, if the electromagnetic force appeared during inflation, the universe would be expanding faster than the speed of light, so all of the universe would be connected. As the electromagnetic field appeared, it would appear all at once and with the same orientation everywhere. No monopoles would be produced. Furthermore, if it should happen that some monopoles had been produced *before* inflation, their overall density would be diluted during expansion. Some might still exist, but now they would be so few and far between that we would not expect to run into one very often.

Calculations show that it is possible to choose a temperature sequence for inflation in which all the monopoles are lost and do not return. This is essential because the supercooled state is followed by a reheating state that leads to the Big Bang, which in turn is followed by more normal cooling. In some scenarios, the reheating unbreaks the symmetry, which later breaks again when the second cooling starts. In that case, the monopoles that were eliminated by inflation return. With the right choice of temperature, however, this will not happen. You may wonder why we should be able to choose the temperature of the inflationary period in our universe. I will get back to that shortly.

The second problem that the inflationary universe solves is linked to the same causes as the monopole problem and has essentially the same solution. The observable universe is about 24 to 30 billion light-years across. Our best estimate for the age of the universe is that it is less than 20 billion years old; that is, the Big Bang occurred less than 20 billion years ago. Philosophically minded cosmologists noted that, since the speed of light is the limit on how fast information can be communicated, one side of the universe was out of touch with the other. Furthermore, working backwards in time to the Big Bang, it became apparent that this was always the case. It did not seem reasonable that parts of the universe that had never been in touch with each other in any way would have evolved in exactly the same way; yet, the whole universe looks the same to us. Furthermore, the residual energy from the Big Bang is almost exactly the same in every direction. If parts of the universe had never communicated with each other, then this residual energy should show considerable variation depending upon in which direction you looked. In other words, there is no reason why the Big Bang should be perfectly symmetrical if one part of the explosion did not know what was going on with the other part.

This problem does not exist with the inflationary universe. During the period of inflation, the universe was expanding faster than the speed of light, so parts that were in touch with each other at the beginning could quickly be dispersed. In this way, although they would never be in communication again, they would share the same origin, allowing the same physical laws to be present from the beginning, which would produce similar evolutionary patterns. The Big Bang would be uniform in all directions, producing the observed symmetrical energy we see today.

The third major problem is that the Big Bang produced almost exactly the amount of expansion necessary to permit the universe to exist. A bit more, and everything would fly off into emptiness; galaxies, stars, planets, and we would not evolve. A bit less, and the universe would come crashing down into itself in a giant black hole. Cosmologists put it this way: The constant omega is 1. Omega is the ratio of the mass density of the universe to the critical mass density needed for the Big Bang to have occurred. The mathematics of the Big Bang make 1 a rather unlikely value for omega. Omega also determines the degree of curvature of space, which results from Einstein's general theory of relativity. When the cosmological constant is 1, the curvature is 0; space is flat. This problem, consequently, is known as the "flatness problem." Some physicists resort to the "anthropic principle" to explain the flatness problem. Roughly speaking, this principle is that if omega were different from 1, the universe would not exist and we would not be there to observe it; we are here—and so omega has to be 1. Many physicists find this to be circular reasoning.

No need for such reasoning. The inflationary universe predicts that *whatever* the initial conditions of the universe, when the period of inflation ends, omega will have to be 1. This is because the mass density of the universe controls the curvature of space during the inflation. As inflation proceeds, no matter what space was like beforehand (if space existed beforehand), it quickly evolves into flatness.

Has the inflationary universe solved all the problems of the cosmologists? Not quite. The original Big Bang theory did not explain the origin of galaxies, clusters of galaxies, and superclusters of galaxies. These are all observed, but there seems to be no good reason for them to exist. The Big Bang would have had to develop in a very specific, and peculiar way for matter to coalesce into the form we see it in today.

Interestingly, one current view of how this organization of the universe has developed finds another symmetry on a truly grand scale. This is known as the sponge theory of the organization of the universe, which contrasts with earlier organization theories known as the island universe theory, the clump theory, the pancake theory, and the bubble theory. While several of these—especially the pancake and bubble theories—have active proponents, the sponge theorists seem to be gaining credence. Briefly stated, the sponge theory says that the universe is organized like a giant bathroom sponge, the kind that was once a living poriferan in the sea. In such a sponge, the solid parts are all connected in one pattern, as one would expect from a formerly living organism. But the holes are also all connected, as they would have to be for sea water (containing the sponge's food) to flow through them. The holes and the solid parts of a sponge are symmetric. That is, if you filled all the holes in a sponge with plaster of Paris and then dissolved the sponge in acid, the resulting cast of the former holes would be just like a sponge. For one thing, the holes in the cast would be all interconnected, and the cast itself would also be interconnected.

In the universe, the sponge is different in that the "solid" parts are not solid. Instead, they are superclusters of galaxies that are all connected into one universal structure. It is regions of high density, where there is a lot of mass, that are connected. The holes, on the other hand, are just like the holes in a sponge. One reason that astronomers think that the universe might have this sponge structure is that it symmetrical. Regions of initial high mass would attract more mass, while regions with little initial mass would give it up. This still does not explain why, in the beginning, some regions had high mass and others had low mass.

The inflationary universe does provide some help. The end of the period of inflation is caused by a phenomenon known as the quantum decay of a false vacuum. Quantum vacuums are filled with energy and a false vacuum is one in which the energy is so high that the state is unstable. It is like an atom that has been lifted to a higher energy state by absorbing a photon, which eventually must decay to a lower state by emitting a photon. It can be shown that when a false vacuum decays, it does not do so all at once. Therefore, fluctuations in the amount of matter will appear during the Big Bang. These fluctuations are the right type to eventually lead to the observed distribution of galaxies. Unfortunately, they turn out to be—in almost everyone's calculations—of the wrong magnitude. Using the SU(5) GUT as the basis of calculations gives the wrong result by a factor of about 100,000—too big to overlook. Other theories of particle physics give different results, and some physicists think that this may be one good way to determine which of the various theories is correct. For example, it appears that supersymmetric SU(5) combined with $N = 1$ supergravity can produce fluctuations of the right size. This depends on getting the initial conditions just right, however. Such a conclusion is contrary to the spirit of the inflationary universe, which otherwise seems to work without any initial conditions whatsoever.

Let's return to the history of the universe, as assembled from various theories.

The History of the Universe (Part 2)

In the beginning there was perfect symmetry.

There is one field with no spin. The field is a quantum field, so there can be random fluctuations of energy in it, even in a vacuum. The field is supercooled so a random fluctuation can start a phase change of the whole field.

A random fluctuation of energy causes the field to start doubling in size every 10^{-34} second. This is the inflationary phase. There is perfect supersymmetry, so particles have no mass, and fermions can change into bosons and back. During this phase supersymmetry breaks, at about 10^{-44} second after the expansion begins, giving much more mass to the superparticles than to the particles. Thus, at this time, gravity separates from the other forces, since mass produces the gravitational force. By 10^{-33} second, the inflationary phase is all over.

During the early universe, at times that have been variously calculated, symmetry breaking also results in the collapse of $N = 8$ supergravity in 10 or 11 dimensions into $N = 1$ supergravity in 4 apparent dimensions. The other 6 or 7 dimensions become compact, making them essentially unobservable. In some versions, this does not happen all at once. Each time a dimension becomes compact, a new internal symmetry is created. These internal symmetries control the fields that are the fundamental reality of our present universe.

The universe reheats as a result of the collapse of the original field. Energy pours from the field into the universe, producing the SU(5) universe, one filled with X particles as well as leptons, quarks, and gluons. Leptons and quarks can change freely into each other. There are now two forces, gravity and the GUT force. Most of the superparticles are gone by now. The amount of matter and antimatter created by the field's decay is equal, but the X particles decay preferentially into matter. The reheating has caused the Big Bang.

Expansion caused by the Big Bang now causes cooling in the universe. The SU(5) universe is also unstable, and SU(5) symmetry breaks, producing an SU(3) × SU(2) × U(1) universe, but one in which the electroweak force exists. There are now real W and Z particles on the scene. By about 10^{-10} second after the start of the universe, however, symmetry is broken again and the electromagnetic force separates from the weak force. The W and Z particles decay.

At this point, there are four forces: gravity, electromagnetism, the weak force, and color. By about 10^{-4} second, the color force has grown strong enough to create hadrons, some of which quickly decay into protons and neutrons. The color force is replaced by the strong force.

Now there are protons, neutrons, and electrons. They begin to come together to form hydrogen and helium. Because of fluctuations in the decay of the original quantum field, there is more hydrogen and helium in some places than others. Gravity begins to form clouds of gas that can condense into stars. High mass attracts mass, low mass gives up mass, and the universe starts to turn into a giant sponge.

About 10^{16} seconds later, people on Earth begin building giant particle accelerators in hopes of restoring some of the symmetry of the early universe for a fraction of a second in hopes of verifying that some of what I have just told you is true.

Our Own Special Universe

Let's return to the question of why one should be allowed to choose the temperature at which inflation occurred. Inflation does not happen like a balloon being blown up very fast. Instead, it is more like the production of foam from a can of shaving soap. The ball of lather expands in your hand because it is made of tiny bubbles that are each expanding.

Similar bubbles of inflation also existed in the early universe. The bubbles were not connected with each other in any way. They never did become connected, for inflation proceeds so fast that the outer edge of each bubble is moving faster than the speed of light. Within a bubble, everything that was connected before the inflation remains connected as a result of the speed of inflation. It is not necessary, or even likely, that the evolution of one bubble proceeds along the same lines as the evolution of any other bubble.

Now comes the hard part. Each one of these bubbles became its own universe. We cannot see these other universes because they are not connected with ours in any way. Neither electromagnetic radiation nor gravity, to name the two long-range forces in our universe, can affect our universe from any of these other universes, since each one of the universes was expanding faster than the speed of light during the inflationary period.

Still, we can do more than speculate about what happened in these other universes. Mathematics has the power to reveal what happens under different conditions.

For example, in some bubbles the conditions for a graceful exit may never develop. Those bubbles are still inflating.

In other bubbles, the temperature at which supersymmetry breaks or at which SU(5) breaks is reached at a different stage of inflation or subsequent Big Bang expansion. Such a universe might be filled with magnetic monopoles, which would cause it to self destruct.

We don't know why 6 or 7 dimensions would have become compact and 4 dimensions would not. Therefore, other universes could have different dimensions. Or, there could be a universe with 4 dimensions, but 4 different dimensions than the ones we are used to. Such a universe is very hard to imagine, but it could exist.

Most intriguing, it turns out that the mathematics of SU(5) symmetry breaking permits several different options. In our universe, SU(5) broke into SU(3) × SU(2) × U(1). This does not necessarily have to be the case. It can be shown that SU(5) could also break into SU(4) × U(1). Such a universe would have very different physical laws than ours does. In a sense, it would have more symmetry than we have. There are other possibilities with less symmetry.

Now we come to why we can choose a temperature at which inflation occurs. We are pretty sure that the universe we live in is, at this time, an SU(3) × SU(2) × U(1) universe. Therefore, we can choose a temperature that will produce such a universe. Other temperatures can exist in other inflationary bubbles, but not in this one.

At first, this argument may seem to be as circular as the one that says omega must be 1 because otherwise we would not exist to measure it. I don't think that the two ideas are comparable. The "fine tuning" of the temperature of inflation is possible because there are a large number of different universes. We can choose a temperature because we can observe the consequences. We are not saying that a particular temperature had to exist so that it would produce a single universe that fortuitously has the right conditions for human beings to evolve. We are saying that of the universes produced by inflation, the one we live in must have developed at a temperature consistent with what we know of this particular universe.

Perhaps another universe would be better than this one in some ways. I don't know what life in an SU(4) × U(1) universe with 5 observable dimensions would be like, assuming that it could exist. Years ago, e. e. cummings wrote, "there's a helluva good universe next door/let's go." Maybe he knew something that we don't.

In our universe at least we know that symmetry is at the heart of understanding. The symmetries that still exist in the universe give us the conservation laws that govern physics. The higher symmetry of group theory explains why the ultimate (as far as we now know) particles are as they are and behave as they do. Supersymmetry may explain the mysteries of gravity. As symmetries break from the initial perfect symmetry, our universe obtains the mass and the forces that make it work. The universe becomes one in which people can evolve so that they can notice that one thing looks just like the other.

Selected Further Reading

Coxeter, H. S. M. *Introduction to Geometry.* 2d. ed. New York, John Wiley & Sons, Inc., 1969. Although this is essentially a college textbook, it is accessible to the general reader who is sufficiently interested in mathematics and it presents with great clarity and some wit the basic geometry of symmetry, groups, and transformations, along with much else that is fascinating.

Crease, Robert P., and Charles C. Mann. *The Second Creation.* New York, Macmillan, 1986. As a result of extensive interviews, this book provides the inside story of much of the recent history of particle physics.

Gardner, Martin. *The Ambidextrous Universe: Mirror Asymmetry and Time-Reversed Worlds.* 2d. rev. ed. New York, Charles Scribner's Sons, 1979. This is the grandfather of all post-Yang-and-Lee popular books on the subject and a real delight. A third edition is due soon.

Grünbaum, Branko, and G. C. Shephard. *Tilings and Patterns.* New York, W.H. Freeman and Company, 1987. This is a survey of tesselations and related subjects by two noted professors of mathematics.

Mandelbrot, Benoit B. *The Fractal Geometry of Nature.* New York, W.H. Freeman and Company, 1983. Although frequently difficult to follow, this remains the Bible of fractal mathematics. The illustrations alone are worth checking the book out.

Schwinger, Julian. *Einstein's Legacy.* New York, W.H. Freeman and Company, 1986. This is a readable account of the theories of relativity (although laced with some of the real mathematics) by one of today's great physicists.

Shapiro, Robert. *Origins: A Skeptic's Guide to the Creation of Life on Earth.* New York, Summit Books, 1986. This book presents and criticizes in detail all of the various theories of how life began, with considerable concern for the problem of the asymmetry of living molecules.

Springer, Sally P., and Georg Deutsch. *Left Brain, Right Brain.* Rev. ed. New York, W.H. Freeman and Company, 1985. This is the definitive essay on all the implications of mental asymmetry between the halves of the brain and between men and women, although new research continues to change the way we think about thinking.

Weaver, Jefferson Hane. *The World of Physics.* New York, Simon and Schuster, 1987. Volumes II and III of this anthology contain a number of essays on topics from *Reality's Mirror,* often by the physicist who originally developed the ideas.

Zee, A. *Fearful Symmetry.* New York, Macmillan, 1986. Covers about the same ground as the physics portion of *Reality's Mirror,* but at somewhat greater length.

Index